A Hybrid Data-Model and AI-Driven Approach for Structural Monitoring in Hazardous Construction

Qiang Li · Peixuan Wang · Bawar Iftikhar ·
Yan Jiang · Mingfeng Huang

A Hybrid Data-Model and AI-Driven Approach for Structural Monitoring in Hazardous Construction

Springer

Author
See next page

ISBN 978-981-95-8687-5 ISBN 978-981-95-8688-2 (eBook)
https://doi.org/10.1007/978-981-95-8688-2

This work was supported by Ningbo Youth Science and Technology Innovation Leading Talent Program (2025QL050), Ningbo Public Welfare Research Program (2023S004, 2024S005, 2025S023), National Natural Science Foundation of China (52508586), Natural Science Foundation of Zhejiang Province (LY24E080012), Ningbo Key R&D Program (2023Z221, 2024Z287, 2025Z037) and Ningbo International Sci-tech Cooperation Project (2024H019).

This Springer imprint is published by the registered company Springer Nature Singapore Pte Ltd.
The registered company address is: 152 Beach Road, #21-01/04 Gateway East, Singapore 189721, Singapore

Qiang Li
School of Civil Engineering
NingboTech University
Ningbo, Zhejiang, China

Bawar Iftikhar
School of Civil Engineering
NingboTech University
Ningbo, Zhejiang, China

College of Civil Engineering
and Architecture
Zhejiang University
Hangzhou, China

Mingfeng Huang
College of Civil Engineering
and Architecture
Zhejiang University
Hangzhou, China

School of Civil Engineering
and Architecture
Guangxi University
Nanning, China

Peixuan Wang
Ningbo Construction Engineering Safety
and Quality Management Service Center
Ningbo, China

Yan Jiang
School of Civil Engineering
Chongqing Jiaotong University
Chongqing, China

Acknowledgements The work described in this study was supported by the National Natural Science Foundation of China (52508586); Zhejiang Provincial Natural Science Foundation of China (LY24E080012); Ningbo Youth Science and Technology Innovation Leading Talent Program (2025QL050); Ningbo Key R&D Program (2023Z221, 2024Z287, 2025Z037); Ningbo Public Welfare Research Program (2023S004, 2024S005, 2025S023); Ningbo International Sci-tech Cooperation Project (2024H019).

Competing Interests The authors have no competing interests to declare that are relevant to the content of this manuscript.

Contents

Chapter 1
Introduction

Abstract The background and significance of research on hybrid data-model-driven safety monitoring and early warning technologies for hazardous engineering structures are presented. A comprehensive review and summary of the current domestic and international research status in structural safety monitoring, safety management, risk prediction, and hybrid data-model-driven methods for hazardous works are provided. The research content and chapter organization of this study are introduced.

Keywords Structural safety monitoring · Artificial intelligence · Risk prediction · Hybrid data-model-driven methods

1.1 Research Background and Significance

In recent years, with the in-depth implementation of national strategies such as "New-Type Urbanization" and "Strengthening the Nation through Transportation," China has continuously increased investment in and construction of municipal infrastructure, leading to the emergence of a large number of major municipal projects, including large-scale bridges, interchanges, and elevated road systems. These large municipal projects are generally characterized by high clearance, large spans, long continuous sections, and irregular structural forms, which have resulted in the increasingly widespread use of hazardous sub-item engineering works (hereinafter referred to as major hazardous works) during the construction stage.

According to Order No. 37 of the Ministry of Housing and Urban–Rural Development of the People's Republic of China, Regulations on the Safety Management of Hazardous Sub-item Engineering Works, major hazardous works refer to sub-item construction works in building and municipal infrastructure projects that are prone to causing mass casualties or significant economic losses during construction. Typical examples include tower cranes, high formwork support systems, and deep foundation pits. Furthermore, the document Notice of the General Office of the Ministry of Housing and Urban–Rural Development on Issues Related to the Implementation of the Regulations on the Safety Management of Hazardous Sub-item Engineering

Q. Li et al., *A Hybrid Data-Model and AI-Driven Approach for Structural Monitoring in Hazardous Construction*, https://doi.org/10.1007/978-981-95-8688-2_1

Works (Jianban Zhi [2018] No. 31) specifies that major hazardous works include, but are not limited to: concrete formwork support systems with a erection height of 5 m or more, a span of 10 m or more, a total construction load of 10 kN/m^2 or more, a concentrated line load of 15 kN/m or more, or structures whose height exceeds the horizontal projection width of the support and consist of relatively independent members without effective connections; as well as hoisting operations involving the installation and dismantling of lifting machinery, the use of unconventional lifting equipment or methods, and single lifting weights of 10 kN or more.

Influenced by multiple factors, major hazardous engineering structures often exhibit low lateral stiffness and poor overall stability during construction, making instability and collapse accidents relatively frequent. As indispensable heavy equipment on construction sites, tower cranes, owing to their tall and flexible structural characteristics, face particularly high collapse risks under extreme weather conditions. In recent years, frequent "gray swan" typhoon events along China's southeastern coastal regions have brought strong winds and heavy rainfall, posing severe threats to tower crane safety at coastal construction sites, as illustrated in Fig. 1. According to incomplete statistics, in September 2016, Typhoon Meranti made landfall in Xiang'an District, Xiamen, causing the collapse of 79 tower cranes and damage to 30 tower cranes at construction sites across the city. In August 2017, Typhoon Hato struck Jinwan District, Zhuhai, resulting in varying degrees of damage to 137 tower cranes. In July 2021, Typhoon In-fa made landfall in Zhejiang Province with maximum wind speeds reaching Level 13 near its center, posing a significant threat to the stability of tower cranes in coastal areas. In 2024, during the passage of the super typhoon Yagi across Haikou, Guangdong, and surrounding regions, numerous tower cranes were broken by strong winds and collapsed onto nearby residential buildings, causing wall cracking and plaster detachment to varying degrees and seriously endangering public safety.

In practical engineering projects, as construction height increases, the self-weight of support systems grows, deformation intensifies, and structural stability decreases. Once a high formwork support system experiences sudden load variations during concrete pouring, ensuring its safety becomes increasingly difficult. As temporary structures, high formwork systems are often subject to less stringent standards in erection, dismantling, and full-process management compared with permanent structural construction, which has led to frequent collapse accidents, as shown in Fig. 2. For example, in 2017, improper concrete pouring procedures at the Xianshan Peony Expo Park Water Amusement Project in Macheng, Hubei Province, resulted in 9 fatalities and 8 injuries. In 2019, the collapse of a formwork support system due to noncompliant structural parameters at the Garden Home Furnishing Market construction project in Nanma Town, Dongyang City, caused 5 deaths and 5 injuries. In 2020, a high formwork collapse occurred at a tourism development project in Wuhan due to improper erection and concentrated loading, resulting in 6 deaths and 5 injuries. In 2023, a high formwork support structure in a railway logistics project in Shanxi Province collapsed because of excessive spacing between vertical members and axial compression overload, leading to 7 fatalities.

The above accident cases repeatedly demonstrate that once major hazardous engineering structures lose stiffness and stability, they often progress to overall collapse within minutes or even seconds, leaving an extremely limited escape window for on-site personnel. Therefore, real-time and effective safety monitoring and intelligent early warning for major hazardous works are critical means to achieve predictive maintenance and ensure long-term structural safety. However, traditional monitoring methods largely rely on manual inspections and simple sensing technologies, which suffer from limitations such as restricted monitoring coverage, insufficient data processing capability, and delayed warning responses. The rapid development of technologies such as the Internet of Things (IoT) and artificial intelligence has provided new technical approaches for safety monitoring and early warning of major hazardous engineering structures. By integrating data analysis with physical model simulation-namely, hybrid data-model-driven techniques-the structural safety state can be assessed more comprehensively and accurately, enabling early warning and intelligent decision-making.

In summary, to ensure the safety and reliable operation of major hazardous engineering structures during construction, it is necessary to leverage advanced sensing technologies and artificial intelligence algorithms to establish a hybrid data-model-driven safety monitoring and early warning system for major hazardous works, thereby improving the overall technical level of structural safety monitoring and early warning. The successful implementation of this research can provide robust technical support for safety monitoring and early warning of major hazardous engineering structures and is of great significance for enhancing safety management practices, improving the accuracy and reliability of early warning systems, preventing and reducing safety accidents, and ultimately safeguarding human life.

1.2 Review of Related Work

1.2.1 Safety Monitoring Technologies for Major Hazardous and Complex Engineering Structures

Structural safety monitoring technology has long been a research focus in the field of civil engineering. In recent years, with the rapid development of sensors, the Internet of Things (IoT), and computer technologies, research on structural safety monitoring has exhibited a trend toward multi-technology integration, intelligence, and systematization, with the core objective of preventing potential safety accidents through real-time monitoring and data analysis.

In the area of safety monitoring for major hazardous engineering structures, numerous scholars have developed advanced IoT-based sensing and monitoring technologies [1–4]. These technologies deploy various sensors on structures-such as accelerometers, load sensors, distance sensors, tilt sensors, and displacement sensors-to collect structural parameters in real time, including acceleration, lifting load, working radius, inclination, and displacement. For tower crane structures, Wu et al. [5] developed a blockchain-based framework in which an IoT system (including load sensors, distance sensors, tilt sensors, and unmanned aerial vehicles) was employed to semi-automatically collect relevant data, enabling real-time monitoring of crane group safety risks and identification of potential on-site hazards. Zhong et al. [6] proposed an IoT-based safety management system for tower crane clusters, which acquires real-time crane status through sensors measuring trolley horizontal and vertical positions, jib and load angles, as well as tower inclination and wind speed, aiming to enhance safety and reduce collision risks during construction activities. Cheng and Teizer [7] installed an ultra-wideband (UWB) sensing infrastructure to track the positions of ground workers and crane hooks, thereby improving operators' ability to identify potentially unsafe crane behaviors. Luo et al. [8] proposed an autonomous construction site safety monitoring system that utilizes various sensors (radio-frequency identification, positioning sensors, accelerometers, and load measurement units) to collect information and monitor crane operating conditions and worker locations in real time. This system can promptly detect unsafe distances between workers and crane loads and issue corresponding warnings, thereby reducing accident occurrence. For high formwork structures, Yuan et al. [9] developed a cyber-physical system (CPS)-based temporary structure monitoring system, which uses different types of sensors (load sensors, switch sensors, accelerometers, and displacement sensors) to collect performance data of scaffolding structures and transmit them in real time to a cloud database for storage and analysis, enabling immediate alarms when potential hazards are detected; Chen et al. [10] developed an intelligent monitoring system for high formwork structures, which improves construction safety by conducting real-time and high-frequency monitoring of structural settlement, inclination, and displacement; Moon et al. [11] established a local ubiquitous sensor network (USN) for real-time data acquisition to monitor the inclination, deflection, and strain of formwork support systems and compare them with limit values to determine whether the systems remain stable within allowable ranges, and subsequently proposed an integrated environment that combines multiple communication technologies to achieve real-time data acquisition, wireless transmission, visualization, and remote monitoring via mobile devices [12, 13], gradually advancing technological development and application depth and improving the efficiency and effectiveness of construction safety monitoring. In summary, monitoring systems based on IoT and sensor technologies can effectively improve construction site safety and productivity by collecting and analyzing structural parameters in real time, thereby reducing accident risks. However, these contact-based sensors require manual climbing during installation, inspection, maintenance, and dismantling, which poses significant operational safety risks. In addition, extensive cable deployment increases potential safety hazards. Therefore, it is necessary to develop a remote, non-contact monitoring and early warning system suitable for complex construction site environments.

In recent years, computer vision technology, as a typical non-contact monitoring method, has been widely applied in the field of structural monitoring [14–17]. A typical computer vision-based measurement method mainly consists of three parts:

camera calibration, target tracking, and three-dimensional reconstruction [18, 19]. This technology is primarily used in studies on structural surface damage identification [20, 21], vibration response monitoring [22], strain monitoring [23], and risk target tracking [24, 25]. In structural monitoring, measurement accuracy has become a major concern. Factors affecting the accuracy of computer vision-based structural vibration measurements can be divided into internal and external factors. Internal factors include image quality, image motion blur, camera calibration algorithms, and target tracking algorithms, while external factors include camera vibration, camera tilt angle, target size, changes in ambient lighting, fog, temperature, and heat haze [26–28]. Most studies focus on verifying whether newly proposed computer vision-based monitoring systems are effective and accurate, while also considering the influence of certain external factors on measurement results. Some researchers have experimentally investigated how external factors affect computer vision monitoring systems and attempted to identify corresponding solutions. For example, Ye et al. [29] studied the effects of illumination, camera elevation angle, and fog on template matching methods and proposed using adaptive algorithms and predictive algorithms to address measurement accuracy issues caused by lighting conditions and vapor-induced target occlusion; Zhou et al. [30] investigated the impact of long-term environmental temperature variations on visual monitoring systems and proposed that temperature-induced measurement errors could be removed from displacement measurements by exploiting the relationship between measurement error and temperature change or by using time–frequency analysis methods.

In the field of monitoring major hazardous engineering structures, computer vision-based methods have received a certain degree of attention [31–33]. In applications to tower crane structures, some researchers installed top-view cameras on crane jibs with the optical axis pointing toward the ground to perform image recognition of hazardous areas, workers, or crane hooks. For example, Yang et al. [34] proposed a novel safety distance monitoring method that collects video data using pan-tilt-zoom (PTZ) cameras and applies Mask R-CNN to identify potential hazards, hooks, and workers, helping crane operators work more effectively and reducing collision accidents; however, such PTZ cameras can only capture the working plane and are highly sensitive to lighting conditions. Wang et al. [35] proposed a real-time three-dimensional spatial map reconstruction method based on images from PTZ cameras mounted on crane tops, which displays workspace boundary lines during crane operations to improve safety and efficiency; Price et al. [36] developed a real-time tower crane monitoring system integrating vision systems and laser scanning technology, in which a monocular camera mounted at the jib tip is used for load tracking and detection of nearby workers, thereby improving safety and efficiency in blind lifting operations. However, the use of top-view cameras mounted on fixtures introduces additional installation complexity, interference, and limited fields of view. To address these issues, some scholars have utilized existing monocular long-range surveillance PTZ cameras for remote monitoring. For example, Wang et al. [37] proposed a framework for real-time recognition and localization of tower cranes and hooks using monocular long-range cameras, which integrates multiple advanced computer vision methods to achieve real-time three-dimensional positioning of tower cranes, including precise monitoring of jib orientation and hook position, thereby supporting safety supervision for construction sites and crane operators and reducing collision accidents; Yang et al. [38] installed PTZ cameras on the rooftops of buildings near construction sites and automatically identified the operational states of tower cranes during construction, such as concrete pouring and non-concrete material transportation, by tracking jib rotation angles and combining site layout information, thereby improving construction efficiency and safety. In applications to high-formwork structures, an intelligent real-time construction monitoring method based on computer vision and finite element analysis was established [39], enabling remote, real-time, and multi-point monitoring of support structures; Feng et al. [40] compared the performance of three algorithms-population-based intelligent digital image correlation (DIC), David Lowe's scale-invariant feature transform (SIFT), and Hessian matrix-based speeded-up robust features (SURF)-in feature detection and matching, and proposed a vision-based monitoring method capable of accurately monitoring displacement and strain of support structures in real time, thereby ensuring construction safety; Jung et al. [41] proposed using image processing techniques to automatically detect deformation of support systems and identify minor changes through small-scale experiments; Luo et al. [42] achieved three-dimensional deformation monitoring of support systems using laser point cloud data by establishing and comparing scaled pipe-axis models at different time points. However, during concrete pouring operations, frequent occlusion of optical paths by workers and machinery often leads to target tracking failure and centroid drift, resulting in distorted monitoring results. To address this issue, a real-time monitoring method for high formwork support systems based on online cameras was proposed to resolve target occlusion challenges in complex environments, and this method has been successfully applied in practical formwork engineering projects [43].

Some innovative technologies have also been preliminarily applied to major hazardous engineering structures. For example, Zhang and Pan [44] proposed the use of virtual reality (VR) technology to allow workers to immersively experience various safety risks that may be encountered during construction; He et al. [45] suggested that augmented reality (AR) technology could overlay real-time data or instructions onto crane control devices, helping operators understand data and interact with equipment, thereby enabling informed and precise decision-making during maintenance and production tasks. However, VR and AR systems involve high costs in hardware, software development, and operator training, and may not be suitable for smaller projects with limited resources. In addition, some researchers have applied unmanned aerial vehicles (UAVs) to structural monitoring [46]; however, due to limited battery life and safety concerns, UAVs cannot remain airborne for long durations.

Although significant progress has been made in real-time monitoring of major hazardous engineering structures, there are still some deficiencies in prediction and early warning. Most existing monitoring systems are limited to data collection and recording and lack reliable early warning capabilities, making it difficult to accurately predict structural response trends. When structures face potential risks, insufficient emergency response time is provided for construction sites, thereby increasing safety risks. Moreover, for high formwork structures, the setting of alarm thresholds in existing monitoring systems is usually based on general safety standards and manufacturers' recommendations. Since each project has unique design characteristics, construction conditions, and service environments, these standards may not be applicable to specific projects. Therefore, the development of scientific and reasonable methods for determining early warning thresholds is of great importance.

1.2.2 Artificial Intelligence in Safety Management and Risk Prediction of Critical Engineering Systems

Machine learning (ML) and deep learning, as core technologies of artificial intelligence, have broad application prospects in the engineering field [47, 48]. Owing to their strong data pattern recognition capabilities and powerful nonlinear modeling abilities, they demonstrate clear advantages in handling complex engineering problems. As a result, a large number of studies have been conducted on safety management and risk prediction for major hazardous engineering structures.

In the area of safety management for major hazardous engineering works, existing studies mainly focus on three aspects: real-time structural monitoring and state assessment, unsafe behavior identification, and operation optimization and path planning [49–54]. For example, in real-time structural monitoring and state assessment, Sakhakarmi et al. [55] used support vector machines (SVMs) to analyze real-time strain data to ensure scaffold safety, achieving prediction accuracy exceeding 98% with limited datasets and demonstrating the potential of ML for real-time scaffold assessment. By leveraging the strong random search capability and global optimization tendency of genetic algorithms, an improved genetic algorithm was developed and integrated with a BP neural network using the inverse of the sum of squared errors as the fitness function [56]. This approach effectively predicted and identified the operational states of tower cranes for fault diagnosis, demonstrating the effectiveness of combining optimization algorithms with neural networks for state assessment. In terms of unsafe behavior identification, Ding et al. [54] developed an unsupervised learning model based on generative adversarial networks (GANs), incorporating five pseudo-anomaly synthesizers to detect unsafe behaviors of construction workers; Jiang et al. [57] proposed a transfer learning-based recognition framework to identify unsafe tower crane hoisting behaviors, particularly inclined lifting, sudden braking, and sudden unloading, with results showing that the transfer learning model achieved a recognition accuracy of 76.74%, outperforming other methods. Current studies mostly focus on identifying single behaviors or individual components; however, in real construction environments, multiple unsafe behaviors are often interrelated. In the area of operation optimization and path planning, Yin et al. [58] proposed a cooperative co-evolutionary genetic algorithm (CCGA) to optimize the service scheduling problem of overlapping tower cranes and compared its performance with traditional genetic algorithms (TGA) and nearest neighbor algorithms (NNA), showing that CCGA outperformed the others in terms of optimal completion time and required

CPU time; Kim et al. [59] adopted a deep learning-based deep global registration (DGR) algorithm to determine relative transformations between consecutive point clouds in scaffold structures.

In terms of risk prediction, research has mainly focused on structural dynamic response prediction, stability prediction, fault prediction, and collision avoidance prediction [60–67]. For example, Li et al. [64] established a real-time tower crane monitoring system based on IoT technology to achieve long-term monitoring of crane dynamic responses under extreme weather conditions such as typhoons, and, based on on-site measured wind speed data, used BP and RBF ML models to predict the maximum displacement of tower cranes in advance under the strong typhoon "In-fa"; Zhang and Ge [65] applied a trajectory prediction model based on the transformer algorithm to predict the future trajectories of monitored objects within the next 10 s using historical data, helping workers avoid potential collision accidents and improving tower crane construction safety; Wei [66] developed an LSTM neural network-based warning system to predict and prevent scaffold failures under extreme wind conditions, integrating wind speed prediction models, wind force analysis, and scaffold collapse assessment to address safety issues in typhoon-prone regions. Moreover, tower crane safety prediction was predicted by constructing ML models [67] that use load, working radius, rotation angle, wind speed, and height as inputs to predict crane moment variations, and employed a cuckoo algorithm to optimize neural networks to avoid local optima during training, thereby improving prediction accuracy.

In summary, data-driven methods based on artificial intelligence have achieved significant breakthroughs and widespread application in engineering management and monitoring; however, many challenges remain. For instance, although traditional data-driven approaches can achieve good predictive accuracy within the range of available datasets, they heavily depend on the completeness of historical data. When data are incomplete, difficult to obtain, or when samples of extreme events are insufficient, these methods exhibit notable limitations [68]. Therefore, to overcome these shortcomings, it is necessary to improve traditional data-driven approaches in order to enhance the accuracy of structural response prediction under extreme conditions.

1.2.3 State of the Art of Hybrid Data-Model-Driven Methods

In recent years, hybrid data-model-driven methods have become one of the key technologies for ensuring structural safety and improving project management efficiency. In the field of civil engineering, the applications of such methods can generally be classified into two categories. The first category involves using on-site monitoring data to perform real-time calibration, error compensation, and parameter updating of digital models, governing equations, or physical models, thereby enabling model-based representations to better reflect actual operating conditions. The second category focuses on generating a large number of extreme-condition or missing-condition samples through finite element simulations to compensate for insufficient historical data, thereby enabling early warning and prediction of structural responses.

Within hybrid data-model-driven applications, the establishment of digital twin models plays a critical role in structural safety management. Numerous studies have demonstrated that integrating Building Information Modeling (BIM) with IoT technologies can significantly improve efficiency by enabling real-time feedback and visualized management, thereby enhancing project collaboration efficiency and decision-making quality [69–74]. For example, Cheng [72] employed Revit software to construct three-dimensional models and connected physical entities with virtual models via IoT technology to enable real-time updating and prediction of building states and performance; Pan and Zhang [73] proposed a digital twin framework based on IoT, BIM, and data mining to promote intelligent construction management; similarly, Zhang et al. [74] developed a digital twin analysis framework integrating BIM, IoT, and data storage and analysis, which was shown to improve construction quality, efficiency, and safety. However, BIM models primarily describe geometric configurations, spatial locations, and material properties, and are unable to directly simulate structural mechanical behavior or predict structural responses under varying loading conditions.

To address these limitations, some researchers have integrated monitoring data with mechanism-based governing equations, thereby improving simulation accuracy and numerical stability. For example, Gao et al. [75] proposed a mechanism-data hybrid-driven method for predicting drilling energy consumption by establishing power and energy models combined with an LSSVM-based data compensation model and an improved whale optimization algorithm, achieving high-accuracy energy consumption prediction and validating the effectiveness of the hybrid-driven approach through multi-perspective visualization. A hybrid-driven urban drainage flow simulation model was developed in which neural networks were employed as error correctors to calibrate numerical solutions of the one-dimensional Saint–Venant equations, demonstrating that the integration of deep learning not only improves simulation accuracy and numerical stability but also reveals intrinsic relationships between modeling errors and input conditions, thereby overcoming key limitations of traditional mechanism-based models [76]. In another study, a mechanism-data hybrid-driven dynamic model for hunting dampers was established by coupling damper mechanism models with neural network techniques, resulting in enhanced model stability and accelerated convergence speed [77]. Furthermore, a mechanism-data hybrid-driven short-process power model was proposed for steel production enterprises, in which a mechanism-based model was used to capture the coupling characteristics of material and energy flows between successive processes, while data-driven models were applied to construct mappings between non-electrical process variables and power function parameters [78]. Case studies based on measured data from a domestic steel enterprise verified the effectiveness and accuracy of the proposed model. Overall, the integration of mechanism-based models and data-driven methods demonstrates strong potential by providing deeper physical insight and compensating for the shortcomings of traditional mechanism models. However, once structural geometry, damage states, or loading conditions change, mechanism-based models often require re-derivation or substantial modification of governing equations and cannot be updated as efficiently as finite element models through mesh modification and parameter adjustment.

With the advancement of computer technology, finite element models (FEMs), as powerful numerical analysis tools, can provide more accurate descriptions of structural mechanical behavior; however, FEMs are highly dependent on precise material parameters and boundary conditions and are often difficult to adapt to uncertainties in real engineering applications. To overcome these limitations, researchers have increasingly combined monitoring technologies with FEMs to achieve more comprehensive structural analysis and prediction [79–88]. For example, Ghahari et al. [84] updated beam FEM using measured data and trained Gaussian process regression models based on residuals between predicted and measured responses to estimate seismic responses of building floors without installed sensors. A three-dimensional milling finite element model was constructed using ABAQUS, and a Python subroutine based on the Fick wear rate model was developed to accurately simulate the evolution of tool wear, with final wear prediction achieved through a MATLAB-based graphical user interface module [85]. In the domain of lifting machinery and special equipment, a comprehensive strategy combining measured samples, finite element simulation, and cyclic sample iteration was proposed to address challenges related to large-scale load spectrum testing and fatigue life simulation errors, thereby enabling reliable evaluation of the remaining fatigue life of crane metal structures [86]. In the field of formwork support systems, Wang et al. [87] developed a reliability-based limit state design method using advanced finite element techniques and updated statistical data on scaffold resistance and construction loads; Cho et al. [88] developed wireless strain sensing modules for real-time data acquisition, constructed FEM models to characterize scaffold behavior under specific loading conditions, and applied support vector machine (SVM) algorithms for decision-making, achieving an average prediction accuracy of 97.66%. Furthermore, some studies have integrated monitoring technologies, BIM, and FEM to establish digital twin frameworks, enabling prediction and analysis of structural mechanical responses under different loading conditions and significantly improving the efficiency and accuracy of engineering management and safety control. For example, Sun et al. [88, 89] established a digital twin framework integrating FEM, BIM, and IoT technologies to achieve efficient data management and four-dimensional visualization of bridge structures; Jiang et al. [90] combined BIM and FEM to develop virtual models and dynamic analyses, and proposed a digital twin-based framework for human–machine interaction in tower crane construction activities to evaluate the effects of unsafe lifting behaviors on crane stability; Liu et al. [91, 92] investigated digital twin-based safety assessment methods for prestressed steel structures by integrating BIM, FEM, monitoring data, and SVM-based predictive models to estimate structural safety risk levels.

In summary, hybrid data-model-driven methods play a crucial role in the digital transformation of the construction industry. By integrating the complementary strengths of data-driven and model-driven approaches, these methods have been preliminarily applied in several engineering projects; however, their application to response early warning and prediction for tower cranes and high formwork support structures remains limited, highlighting the need for further research and development.

1.2.4 Summary and Research Gaps

At present, researchers both domestically and internationally have carried out extensive studies on safety monitoring and early warning technologies for major hazardous engineering structures. Although significant progress has been made in real-time monitoring of major hazardous works, several issues related to structural early warning and prediction still warrant further investigation, as outlined below.

For tower crane structures:

(1) Existing monitoring systems generally lack trend prediction capability and issue alarm signals only after warning thresholds are exceeded, making it difficult to achieve early warning of structural responses.
(2) Most studies adopt traditional data-driven approaches that rely primarily on historical measured data. In practice, such historical data are often incomplete and may lack extreme-condition samples. Consequently, when sudden extreme operating conditions occur, prediction models may underestimate structural responses due to insufficient learning of extreme scenarios, resulting in inaccurate prediction of extreme responses.

For high formwork support structures:

(1) Most existing IoT-based sensing and early warning systems employ contact-based sensors, which require personnel to climb during installation and maintenance, posing significant operational safety risks.
(2) Research on methods for determining and dynamically updating early warning thresholds for high formwork structures is extremely limited.
(3) Existing monitoring systems generally lack trend prediction capability and issue alarms only after warning thresholds are exceeded, making early warning of structural responses difficult to achieve.

1.3 Research Objectives

This study adopts a hybrid data-model-driven approach by integrating advanced sensing technologies, finite element analysis, and artificial intelligence algorithms to achieve safety monitoring and response early warning for tower cranes and high formwork support structures, thereby ensuring construction safety, reducing accident rates, and protecting life and property. The specific research objectives are as follows:

For tower crane structures:

(1) To propose a tower crane wind-induced dynamic response prediction method based on a CNN-BiLSTM-AdaBoost ML model, enabling the capture of response trends and the provision of early warning.
(2) To develop a hybrid data-model-driven prediction framework for tower crane dynamic responses under extreme typhoon conditions, effectively addressing the lack of extreme samples in measured data and improving the accuracy and reliability of predictions for "gray swan" typhoon events.

For high formwork support structures:

(1) To propose a computer vision-based real-time displacement monitoring scheme for high formwork structures.
(2) To develop a hybrid data-model-driven framework for determining and dynamically updating monitoring and early warning thresholds for high formwork structures, enabling continuous iterative updating of alarm thresholds and improving early warning accuracy.
(3) To propose a CNN-BiLSTM-AdaBoost-based ML approach for short-term displacement response prediction and rapid inversion of upper construction loads for high formwork structures, thereby providing early warning of structural responses.

1.4 Research Framework and Technical Roadmap

This study focuses on tower cranes and high formwork support systems, two representative types of major hazardous engineering structures, and adopts a hybrid data-model-driven approach to achieve structural safety monitoring and early warning. The research mainly addresses the prediction of tower crane dynamic responses under extreme typhoon conditions, the determination and updating of monitoring and early warning thresholds for high formwork structures, and the prediction of structural responses of high formwork systems. To better illustrate the overall research concept of this study and the logical relationships among the chapters, the overall research framework is presented in Fig. 1.1.

The specific chapter arrangement of this study is as follows:

(1) **Introduction**: First, the background and significance of research on hybrid data-model-driven safety monitoring and early warning technologies for major hazardous engineering structures are presented. Then, a comprehensive review and summary of the current domestic and international research status in structural safety monitoring, safety management, risk prediction, and hybrid data-model-driven methods for major hazardous works are provided. Finally, the research content and chapter organization of this study are introduced.
(2) **Hybrid Data-Model-Driven Theory**: This chapter focuses on the principles of representative ML models, including CNN, BiLSTM, and AdaBoost. Based

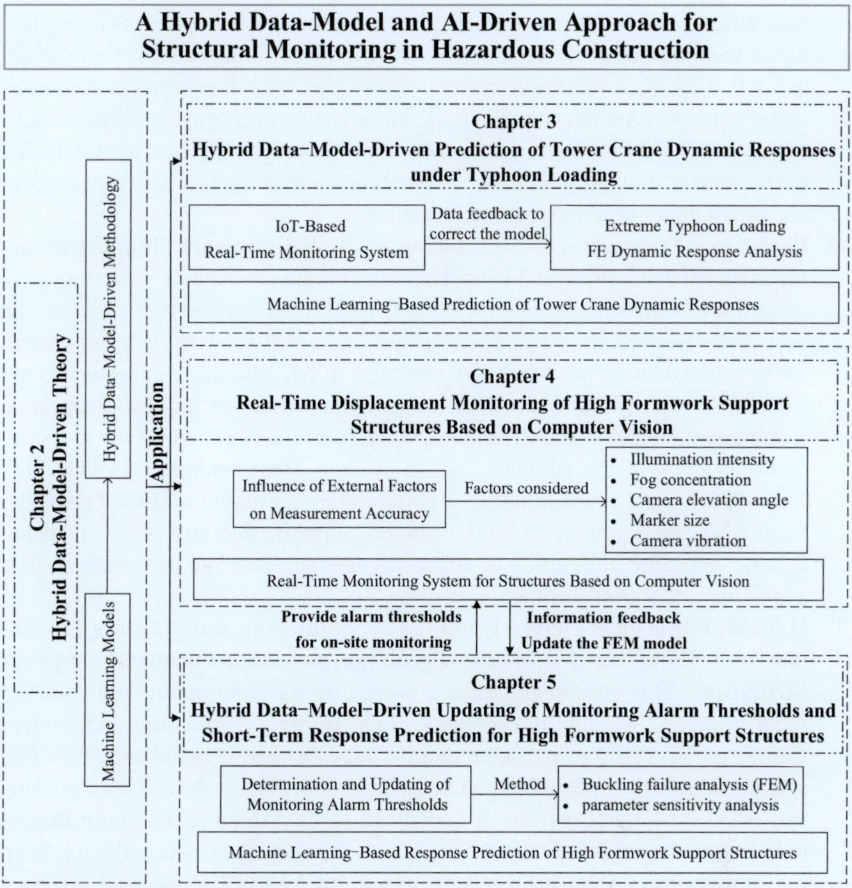

Fig. 1.1 Overall research framework

on an analysis of the advantages and limitations of individual models, a hybrid prediction model based on an adaptive boosting ensemble learning algorithm (CNN-BiLSTM-AdaBoost) is constructed. Subsequently, the overall framework and implementation pathway of the hybrid data-model-driven approach for major hazardous engineering structures are elaborated, providing methodological support for subsequent prediction of tower crane dynamic responses under typhoon conditions, determination and updating of monitoring and warning thresholds for high formwork structures, and prediction of structural responses.

(3) **Hybrid Data-Model-Driven Prediction of Tower Crane Dynamic Responses under Typhoon Loading**: Using the proposed hybrid data-model-driven approach, tower crane response prediction under extreme typhoon conditions is carried out. Taking a tower crane from an actual engineering project as an

example, an IoT-based real-time displacement monitoring system is first established to record tower crane displacement responses during construction. Subsequently, a FEM of the tower crane is developed, and displacement responses under extreme typhoon conditions are simulated to compensate for the lack of extreme samples in measured data. Finally, by integrating measured data and finite element simulation data, ML-based prediction of tower crane displacement responses under typhoon loading is achieved.

(4) **Real-Time Displacement Monitoring of High Formwork Support Structures Based on Computer Vision**: Experiments are conducted to investigate the influence of different external factors on visual measurement accuracy, and the proposed visual monitoring system is applied to real-time displacement monitoring of an actual high formwork structure. First, laboratory experiments are performed to study the effects of various external factors-including illumination, fog, camera elevation angle, target size, and camera vibration-on the measurement accuracy of the computer vision system. Then, taking a practical high formwork project as an example, a computer vision-based real-time displacement monitoring system for high formwork support structures is established to achieve real-time monitoring of structural displacement and deformation, data acquisition, and over-limit early warning.

(5) **Hybrid Data-Model-Driven Updating of Monitoring and Warning Thresholds and Short-Term Response Prediction for High Formwork Support Structures**: This chapter proposes a procedure for determining and updating monitoring and warning thresholds for high formwork structures and realizes short-term structural response prediction. First, a FEM is established based on special design documentation and boundary conditions at the construction site, and buckling failure analysis is conducted to determine initial alarm thresholds for guiding on-site monitoring. Then, the measured data fed back from the monitoring system are used to update the FEM. The updated buckling failure displacement is compared with the initial alarm threshold to evaluate threshold rationality and update the alarm thresholds accordingly. By integrating measured and finite element simulation data, ML-based short-term displacement response prediction and rapid inversion of construction loads for high-formwork structures are achieved, demonstrating the superiority of the CNN-BiLSTM-AdaBoost model for response prediction.

(6) **Conclusions and Future Work**: The main research contributions and innovations of this study are summarized, and directions for future research are discussed.

Chapter 2
Hybrid Data-Model-Driven Theory

Abstract The fundamental principles of three representative machine learning models-Convolutional Neural Networks (CNN), Bidirectional Long Short-Term Memory networks (BiLSTM), and the ensemble learning algorithm AdaBoost-are systematically analyzed. Based on the respective advantages and limitations of individual models, a CNN-BiLSTM-AdaBoost hybrid prediction model is developed. Furthermore, an overall framework and implementation pathway for hybrid data-model-driven monitoring of major hazardous engineering structures are proposed, providing methodological and theoretical support for subsequent prediction of tower crane dynamic responses under typhoon loading, dynamic determination and updating of monitoring and warning thresholds for high formwork support structures, and short-term structural response prediction. In this framework, multi-source monitoring data are acquired in real time by various sensing devices and synchronously transmitted to an early warning platform for integration, visualization, and analysis, while simultaneously driving the dynamic updating of finite element models. Based on measured data and simulation results from the updated models, safety evaluation and response prediction are conducted, and the assessment results and warning information are fed back to the construction site, thereby realizing a closed-loop interaction among the physical structure, the finite element model, and the monitoring and early warning platform.

Keywords Hybrid data-model-driven theory · CNN · BiLSTM · AdaBoost

To address the deficiencies of existing monitoring systems for tower cranes and high formwork support structures in terms of safety monitoring and response, early warning prediction, and to ensure construction safety and reduce accident occurrence, this chapter proposes a hybrid data-model-driven approach for structural monitoring and early warning. This approach integrates the complementary advantages of physical modeling and artificial intelligence algorithms. It not only overcomes the

Q. Li et al., *A Hybrid Data-Model and AI-Driven Approach for Structural Monitoring in Hazardous Construction*, https://doi.org/10.1007/978-981-95-8688-2_2

limitations of traditional physical models, which struggle to adapt to changing real-world operating conditions, but also mitigates the shortcomings of conventional data-driven methods, such as insufficient extreme-event samples and weak generalization capability under extreme conditions.

This chapter discusses both the construction of traditional data-driven models and the principles of hybrid data-model-driven approaches. First, commonly used ML models in the field of safety monitoring and early warning for major hazardous engineering structures-namely Convolutional Neural Networks (CNN) and Bidirectional Long Short-Term Memory networks (BiLSTM)-are reviewed, and their respective advantages and limitations are analyzed. On this basis, the Adaptive Boosting ensemble learning algorithm (AdaBoost) is further introduced to improve the CNN-BiLSTM model, resulting in the development of a CNN-BiLSTM-AdaBoost hybrid model framework. This framework is designed to enhance prediction accuracy under extreme operating conditions and to provide theoretical support for subsequent studies on response monitoring and early warning for tower cranes and high-formwork structures. Subsequently, the overall framework and implementation pathway of the hybrid data-model-driven approach are established, dynamically linking the physical structure, FEM, and monitoring and early warning platform. This integration enables structural monitoring and predictive early warning and provides a systematic research approach for subsequent studies on safety monitoring and early warning of major hazardous engineering structures.

2.1 Fundamentals of Machine Learning Models

Data-driven approaches mainly rely on measured data obtained from practice (including data from sensors, monitoring systems, and historical records) to extract useful information through data analysis. In such approaches, data serve as the core, and ML techniques are used to learn the mapping relationships between inputs and outputs from historical samples. In the field of structural safety monitoring, these methods are commonly applied to state identification, structural monitoring and detection, and short-term prediction.

2.1.1 Convolutional Neural Networks (CNNs)

A convolutional neural network (CNN) is a type of feedforward neural network that employs local connectivity and shared weights to perform high-dimensional mapping of raw data, thereby effectively extracting data features. Its basic architecture consists of an input layer, convolutional layers, pooling layers, fully connected layers, and an output layer, as illustrated in Fig. 2.1.

The convolutional layer is the core component of a CNN and also the layer with relatively high computational cost during model training. It captures local features

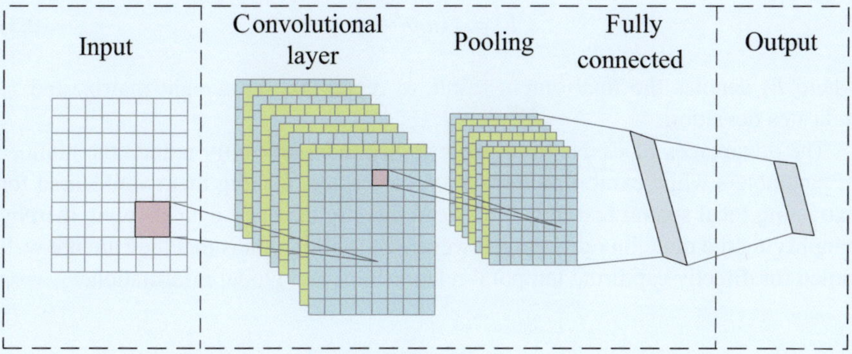

Fig. 2.1 Schematic diagram of the CNN architecture

of the input data through multiple learnable convolution kernels to generate feature maps, which are then passed through nonlinear activation functions (typically the ReLU function) to produce the output of the convolutional layer. In the convolutional layer, the same convolution kernel adopts a shared-weight mechanism, which reduces the number of parameters during training, accelerates model convergence, and enables effective feature extraction from the data. The convolutional layer can be expressed as follows [93].

$$c_i = f(w_i \cdot x_i + b_i) \tag{2.1}$$

In the equation, C_i represents the i-th output feature map of the convolutional layer, f denotes the activation function, w_i is the weight matrix, x_i is the input feature of the convolutional layer, and b_i represents the bias vector.

A CNN usually contains multiple convolutional layers, and each convolutional layer is typically followed by a pooling layer. The pooling layer performs a down-sampling operation on the output of the convolutional layer by compressing the neurons in the input feature maps, thereby reducing model complexity. This process effectively reduces data dimensionality, helps prevent overfitting, and improves the efficiency of feature extraction. Common pooling methods include max pooling and average pooling; in this study, max pooling is adopted. The pooling layer can be expressed as follows:

$$p_i = \max(c_i, c_{i-1}) + \varphi_i \tag{2.2}$$

where p_i denotes the output of the pooling layer, and φ_i represents the bias term.

The fully connected layer is located after the pooling layers and constitutes the high-level part of the network architecture. It integrates and transforms the features extracted by the convolutional and pooling layers to perform final classification or regression tasks and outputs the prediction results. The fully connected layer can be expressed as follows:

$$F_i = f(u_i p_i + \eta_i) \tag{2.3}$$

where F_i denotes the final output result, u_i represents the weight matrix, and η_i indicates deviation.

The advantages of CNNs lie in their ability to significantly reduce the number of parameters while exhibiting translation invariance, making them well suited for extracting local spatial features. CNNs can directly take raw data as input, thereby simplifying the modeling process. However, their limitation is that they are not well suited for directly capturing temporal relationships and global information.

2.1.2 Recurrent Neural Networks (RNNs) and BiLSTM

A recurrent neural network (RNN) is a type of neural network that is well suited for processing time-series data; however, it suffers from the long-term dependency problem, meaning that as the length of the input sequence increases, the model becomes unable to effectively utilize information from earlier parts of the sequence. To address this issue, Hochreiter et al. proposed the Long Short-Term Memory (LSTM) network in 1997. LSTM is a special type of recurrent neural network that introduces memory storage units to overcome the difficulty of learning long-term dependencies in sequences. A typical LSTM unit structure is shown in Fig. 2.2, which mainly consists of a forget gate(f_t), an input gate(i_t), and an output gate(O_t).

The forget gate determines which information from the previous cell state C_{t-1} should be discarded. It can be expressed as follows [94]:

$$f_t = \sigma(W_f[h_{t-1}, I_t] + b_f) \tag{2.4}$$

In the equation, f_t is the output of the forget gate, with values ranging from 0 to 1; σ denotes the Sigmoid activation function; W_f is the weight matrix of the forget

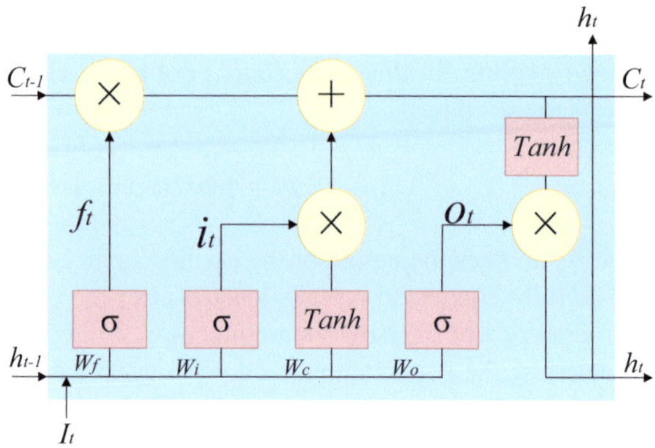

Fig. 2.2 LSTM structure diagram

gate; h_{t-1} represents the hidden state at time t−1; I_t is the input at time t, and b_f is the bias term of the forget gate.

The input gate consists of two parts. The first part, shown in Eq. (2.5), is an activation function layer that determines which information needs to be updated. The second part, shown in Eq. (2.6), is a tanh layer that generates a new candidate value vector and adds it to the cell state. Subsequently, the cell state at time t−1 is updated to the cell state at time t and the update process is given by Eq. (2.7).

$$i_t = \sigma(W_i[h_{t-1}, I_t] + b_i) \tag{2.5}$$

$$\widetilde{C}_t = \tanh(W_c[h_{t-1}, I_t] + b_c) \tag{2.6}$$

$$C_t = f_t * C_{t-1} + i_t * \widetilde{C}_t \tag{2.7}$$

In these equations, i_t denotes the output of the input gate, $\widetilde{C}_t$ is the candidate vector, W_i and W_c are the weight matrices of the input gate, b_i and b_c are the corresponding bias terms, and * represents element-wise multiplication.

The output gate selects important information from the current cell state for output and also consists of two parts. First, the activation function layer determines which neurons should be output. Then, the selected neurons pass through a tanh layer and are multiplied by the output of the activation function layer to obtain the final output, as shown in the following equations:

$$O_t = \sigma(W_o[h_{t-1}, I_t] + b_o) \tag{2.8}$$

$$h_t = O_t * \tanh(C_t) \tag{2.9}$$

where O_t denotes the output of the output gate, h_t is the hidden state at time t, W_o is the weight matrix of the output gate, and b_o is the bias term of the output gate.

The Bidirectional Long Short-Term Memory (BiLSTM) neural network is composed of a forward LSTM network and a backward LSTM network stacked together and represents an improved version of the LSTM architecture. It facilitates the incorporation of both past and future sequence information, thereby improving prediction accuracy [95]. As shown in Fig. 2.3, $X_1, X_2, X_3 \ldots X_t$, and $Y_1, Y_2, Y_3 \ldots Y_t$ denote the input and output data at each time step, respectively; $A_1, A_2, A_3 \ldots A_t$, and $B_1, B_2, B_3 \ldots B_t$ represent the backward and forward LSTM hidden states; and $\omega_1, \omega_2, \omega_3 \ldots \omega_t$ denotes the weight matrices of each layer.

The state at each time step in the model is calculated as shown in Eqs. (2.10) and (2.11), and the final output is computed as shown in Eq. (2.12).

$$A_n = f_1(\omega_2 X_n + \omega_4 A_{n+1}) \tag{2.10}$$

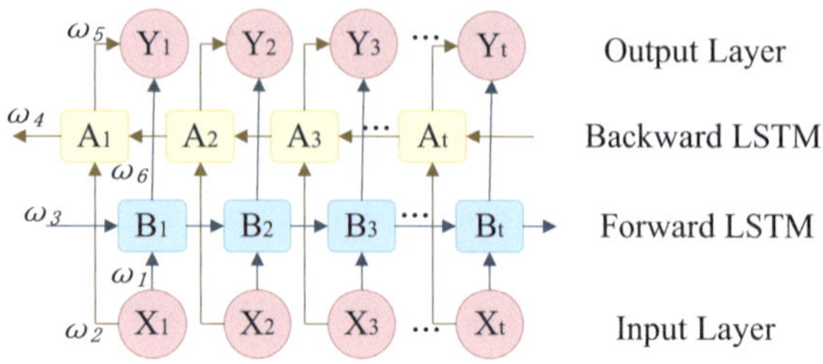

Fig. 2.3 BiLSTM network architecture

$$B_n = f_2(\omega_1 X_n + \omega_3 B_{n-1}) \tag{2.11}$$

$$Y_n = f_3(\omega_5 A_n + \omega_6 B_n) \tag{2.12}$$

where f_1, f_2, f_3 denote the inter-layer activation functions.

The advantages of the BiLSTM model lie in its ability to simultaneously utilize forward and backwards hidden states, enabling it to capture both past and future information in a sequence and thereby enhance temporal representation capability. It can be seamlessly integrated with structures such as CNNs. In addition, the gating mechanism of LSTM (input gate, forget gate, and output gate) can theoretically preserve long-term memory, effectively alleviating the vanishing gradient problem encountered in traditional RNNs. However, the main drawback of BiLSTM is that the number of parameters is approximately twice that of a unidirectional LSTM, which results in higher memory consumption.

2.1.3 AdaBoost Ensemble Learning Algorithm

Boosting is an effective strategy that significantly improves overall predictive performance in regression and classification tasks by iteratively combining multiple weak learners through weighted aggregation, and is also referred to as an "enhancement" or "boosting" learning method. The Adaptive Boosting (AdaBoost) algorithm is one of the most representative boosting algorithms and is characterized by its fast training speed and ease of implementation. The core idea of AdaBoost is to iteratively train multiple weak classifiers and combine them through a weighting strategy to form a stronger classifier with higher accuracy and robustness. In each iteration, AdaBoost computes the loss of each weak classifier and updates the weights of individual training samples, assigning higher weights to misclassified samples so that they receive greater attention in subsequent training rounds. This process enhances the learning effect of the next iteration, as illustrated in Fig. 2.4.

Assume that K denotes the training dataset, which contains a total of N samples, where x_i, y_i represent the input vector and output vector, respectively, as shown in Eq. (2.13). In each iteration, the weight α_i of the weak classifier is calculated based on its classification error e_i as given in Eq. (2.14). Subsequently, the sample weight

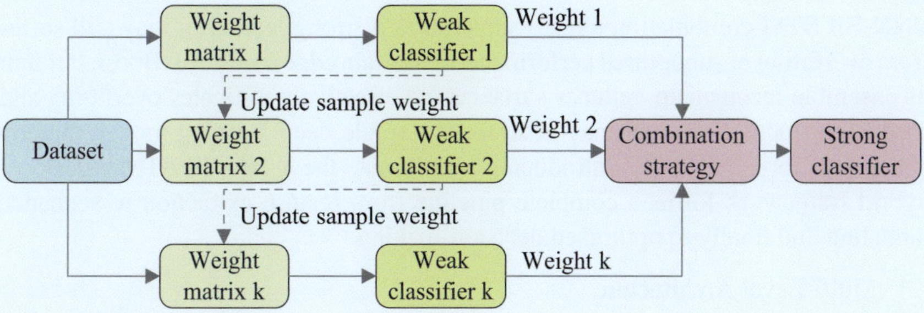

Fig. 2.4 AdaBoost flowchart

distribution D_m is updated according to the performance of the weak classifier on the training dataset, as shown in Eq. (2.15). Finally, the weak classifiers are integrated into a strong classifier $G(x)$ as expressed in Eq. (2.16) [96].

$$K = \{(x_1, y_1), (x_2, y_2), ..., (x_i, y_i), ..., (x_N, y_N)\}, (x_i \in X \in \mathbb{R}^n, y_i \in Y) \quad (2.13)$$

$$\alpha_i = \frac{1}{2}\ln(\frac{1-e_i}{e_i}) \quad (2.14)$$

$$D_{m+1,i} = \frac{D_{m,i}\exp(-\alpha_m y_i G_m(x_i))}{Z_m}, (i = 1, 2, 3, ..., N) \quad (2.15)$$

$$G(x) = sign\left[\sum_{m=1}^{M} \alpha_m G_m(x)\right] \quad (2.16)$$

where Z_m is the normalization factor, which ensures that the sum of the updated sample weights equals 1.

2.1.4 CNN-BiLSTM-AdaBoost Hybrid Algorithm

(1) Advantages of the Hybrid Algorithm

This hybrid algorithm enables complementary strengths and overcomes the limitations of individual models. CNNs exploit parameter sharing, local connectivity, and hierarchical feature extraction to focus on capturing local spatial relationships among data points, achieving efficient learning, particularly in feature extraction tasks. However, CNNs are not well suited for capturing long-range dependencies and temporal patterns. BiLSTM effectively compensates for these shortcomings by treating the feature sequences output by CNNs as time steps and learning contextual information in both forward and backward directions, thereby capturing long-term temporal dependencies and bidirectional contextual relationships. Although the

CNN-BiLSTM combination already constitutes a strong learner, it may still suffer from overfitting or suboptimal performance on certain edge cases. AdaBoost, through its ensemble mechanism, reduces variance and effectively mitigates overfitting and prediction fluctuations that may arise from a single deep learning model, thereby enhancing robustness. By introducing AdaBoost, the CNN-BiLSTM-AdaBoost hybrid framework forms a complete pipeline from feature extraction to sequence modeling and finally to optimized decision-making.

(2) Multi-Level Architecture

The framework consists of an input layer, a CNN layer, a BiLSTM layer, an AdaBoost layer, and an output layer, and has been widely applied in domains such as image recognition and natural language processing. Its core concepts and implementation details are as follows:

- CNN layer: Employs one-dimensional or multi-dimensional convolution kernels to extract local spatial features from the input sequences.
- BiLSTM layer: Dedicated solely to temporal dependency modeling. It receives the high-dimensional feature maps produced by the CNN layer and captures long-term temporal dependencies and bidirectional contextual information through a bidirectional gating mechanism, thereby compensating for CNN's limitations in temporal modeling.
- AdaBoost layer: Responsible for ensemble optimization. It treats multiple CNN-BiLSTM submodels as base learners and constructs a stronger and more robust learner by iteratively adjusting sample weights and combining their prediction results in a weighted manner. This effectively alleviates issues such as susceptibility to local optima and overfitting commonly observed in single models.

(3) Training Procedure

Figure 2.5 illustrates the prediction framework of the hybrid model. During the data preparation stage, raw monitoring data (e.g., wind speed and displacement) are normalized, segmented using a sliding window, and split into training and testing datasets according to a predefined ratio. In the model training stage, the preprocessed training data are first fed into the input layer, after which the CNN convolutional layers extract spatial features and the pooling layers reduce feature dimensionality. The extracted key features are then input into the BiLSTM layer, which captures contextual information in dynamic response sequences in both forward and backward directions. The outputs of the BiLSTM layer are subsequently used as inputs to the AdaBoost module, which integrates multiple weak learners to construct a strong learner, thereby improving predictive performance and reducing overfitting risk. During training, AdaBoost iteratively adjusts the weights of misclassified samples to enhance classification accuracy. After training, new input data passes through CNN-based feature extraction and BiLSTM-based sequence modeling, and predictions are generated via AdaBoost. Finally, evaluation metrics are computed, and visualization analyses-such as predicted versus observed comparisons, fitting curves, and error plots-are produced.

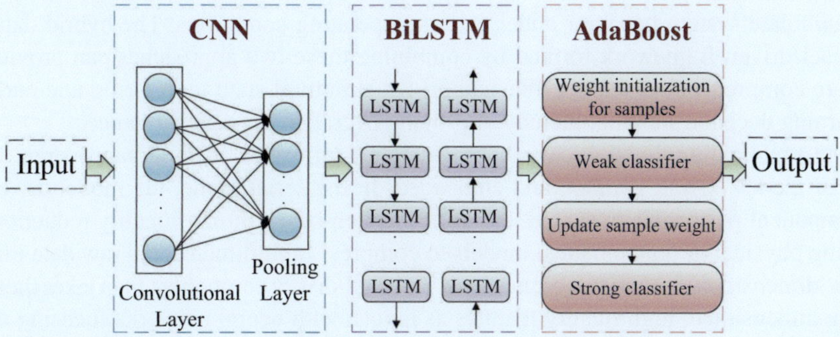

Fig. 2.5 CNN-BiLSTM-AdaBoost prediction framework

To quantitatively evaluate the performance of the models, the root mean square error (RMSE), the coefficient of determination (R^2), and the maximum error are used to assess the established ML models. Among them, RMSE evaluates the overall prediction accuracy and is sensitive to large errors; R^2 represents the correlation between the predicted values and the true values, where a value closer to 1 indicates a better fit and higher model performance; and the maximum error evaluates reliability under the worst-case scenario. The evaluation metrics are expressed as follows:

$$RMSE = \sqrt{\frac{1}{n}\sum_{i=1}^{n}(y_i - \hat{y}_i)^2} \tag{2.17}$$

$$R^2 = 1 - \frac{\sum_{i=1}^{n}(y_i - \hat{y}_i)^2}{\sum_{i=1}^{n}(y_i - \overline{y})^2} \tag{2.18}$$

$$MaxError = \max\left(\left|y_i - \hat{y}_i\right|\right) \tag{2.19}$$

where y_i denotes the true value, $\hat{y}_i$ represents the predicted value, and n is the number of samples.

2.2 Hybrid Data-Model-Driven Methodology

Safety monitoring and early warning of major hazardous engineering structures must address challenges posed by complex operating conditions and extreme environments. In this context, the hybrid data-model-driven method demonstrates significant potential by integrating measured data with physical models to achieve complementary advantages. The data-driven approach, based on ML and historical measured data, is used to mine relationships among variables and predict future trends, while the model-driven approach relies on physical equations and mathematical modeling to simulate system behavior under extreme operating conditions. The hybrid data-model-driven framework formed by combining these two approaches can provide more comprehensive and reliable support for structural state monitoring and early warning decision-making, thereby improving overall system performance.

In practical applications, the hybrid data-model-driven method can operate in three modes: serial, parallel, and embedded. In the serial mode, the model-driven component performs physics-based or mechanism-based dimensionality reduction, using physical or mathematical models to compress high-dimensional raw data into low-dimensional feature representations. The data-driven component then takes these low-dimensional, high-quality features as inputs, with neural networks focusing on residual learning or fine mapping. In the parallel mode, the model-driven and data-driven approaches operate independently and simultaneously, each generating its own prediction results. A fusion mechanism-such as weighted averaging or ensemble voting-is then used to integrate the outputs into a final prediction, which can significantly improve prediction accuracy. However, this mode requires two computational pipelines to run in parallel, resulting in increased computational resource consumption. In the embedded mode, data-driven and model-driven approaches are deeply integrated, with the output of one serving as the input of the other, allowing mutual complementarity and forming an organic hybrid system. For example, Sun et al. [89] achieved bidirectional data flow and real-time interaction between physical entities and virtual models by integrating data-driven and model-driven methods, enabling full life-cycle management and health monitoring of bridge structures.

The hybrid data-model-driven framework for major hazardous engineering structures proposed in this study is illustrated in Fig. 2.6. This framework establishes a closed-loop interaction system among the physical entities, the FEM, and the monitoring and early warning platform. In this system, the monitored high formwork support structures and tower cranes serve as the physical entities. The sensing layer consists of computer vision monitoring systems, IoT sensors, and meteorological monitoring devices, which continuously and in real time collect multi-source data from the physical world, such as displacement, acceleration, wind speed, and video images. The measured data are first transmitted to the monitoring and early warning platform for data integration and visualization and are simultaneously incorporated into the FEM to achieve dynamic model updating through data fusion and expansion. Acting as a carrier of physical mechanisms, the FEM supports simulation analysis, load inversion, and damage identification. Finally, based on the above analyses, the monitoring and early warning platform performs safety assessment and diagnosis and feeds early warning and prediction information back to the actual structure.

The physical entity is the foundation of the hybrid-driven framework, serving as the fundamental basis for constructing the digital model and as the ultimate object that the framework is designed to support. The current state of major hazardous engineering structures is inherently unknown, and throughout their life cycle, these structures are affected by multiple factors such as performance degradation, routine maintenance, damage, and repair, necessitating continuous assessment. Specifically,

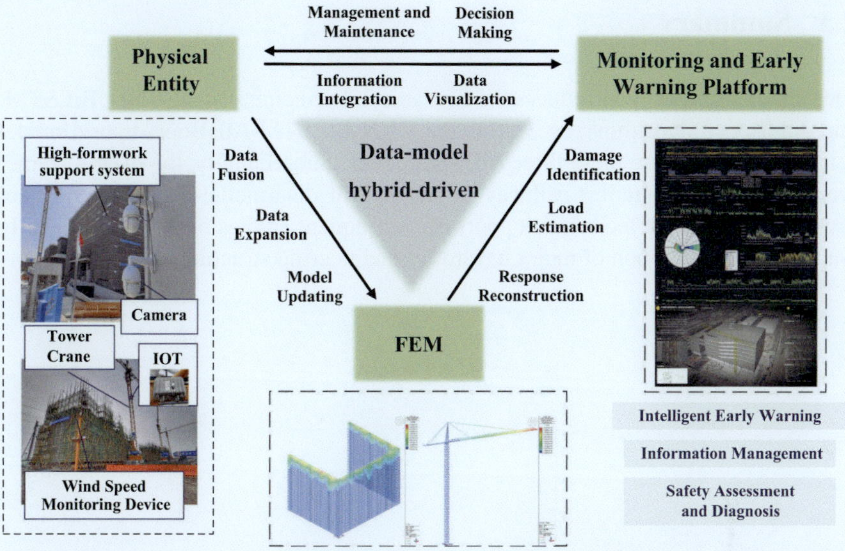

Fig. 2.6 Hybrid data-model driven framework for critical and major engineering structural data

structures are subjected to influences such as environmental conditions, loads, earthquakes, and typhoons, which may cause gradual or sudden deterioration of structural health, making real-time monitoring essential.

The finite element method (FEM) is employed to simulate and analyze the mechanical behavior of real major hazardous engineering structures. By utilizing sparse measured response data, FEM models can inversely estimate the loads acting on the entire structure and simulate global structural responses under different loading conditions, extreme events, or hypothetical damage scenarios, thereby compensating for the spatial limitations of measured data and the scarcity of data under extreme conditions. In addition, when unexplained discrepancies arise between measured values and theoretical predictions, potential structural damage can be inferred.

The monitoring and early warning platform functions as a data integration module that aggregates multi-source information, including sensor data, structural responses, load distributions, and damage conditions. Furthermore, the platform dynamically visualizes three-dimensional structural response results obtained from finite element analyses, supporting informed decision-making by managers, such as determining optimal maintenance timing and optimizing maintenance strategies. Through the hybrid-driven approach, these data are analyzed and predicted using data-driven methods and subsequently managed through visualization within the platform. When monitoring data exhibits anomalies or exceeds safety thresholds, the system immediately issues warnings to managers, enabling timely intervention. Meanwhile, the system dynamically updates the FEM using real-time on-site data during the feedback process, ensuring consistency between the model and the physical entity.

2.3 Summary

This chapter primarily introduces the fundamental principles of the CNN, BiLSTM, and AdaBoost algorithms, as well as the CNN-BiLSTM-AdaBoost hybrid model formed by integrating these three ML techniques. Subsequently, the overall framework and implementation pathway of the hybrid data-model-driven method are presented, providing technical support for subsequent studies on monitoring, early warning, and prediction of major hazardous engineering structures.

Chapter 3
Hybrid Data-Model-Driven Prediction of Tower Crane Dynamic Responses Under Typhoon Loading

Abstract A hybrid data-model-driven method is proposed for predicting tower crane displacement responses under extreme typhoon conditions. Using a tower crane from an actual engineering project as a case study, an IoT-based real-time monitoring system is established to obtain displacement response data throughout the construction period. In parallel, a finite element model is developed to simulate the structural displacement of the tower crane under extreme wind loads, compensating for the scarcity of extreme-condition samples in the monitoring data. Based on the CNN-BiLSTM-AdaBoost algorithm, both purely data-driven and hybrid data-model-driven displacement predictions are conducted for normal wind conditions and extreme typhoon conditions. The results indicate that purely data-driven methods exhibit limited predictive capability under extreme conditions and tend to underestimate structural responses. In contrast, the hybrid data-model-driven approach effectively integrates measured data with finite element simulation results, significantly improving the accuracy and reliability of displacement predictions under extreme typhoon loading.

Keywords Tower crane · Response prediction · CNN-BiLSTM-AdaBoost · Extreme typhoon

With global climate change, the frequency of "gray swan" typhoon events (i.e., low-probability but high-impact events) in the southeastern coastal regions of China has continued to increase, posing a serious risk of collapse to tower cranes at coastal construction sites. Real-time monitoring and response prediction of tower cranes can enable timely identification of structural displacement trends, provide sufficient time for on-site emergency response, and reduce accident risks. However, traditional data-driven methods, which mainly rely on historical measured data, exhibit inherent limitations when confronted with extreme operating conditions. In practice, historical datasets often lack samples corresponding to extreme conditions. As a result, prediction models may underestimate structural responses due to insufficient learning of extreme scenarios, making it difficult to provide accurate response predictions. For example, in "gray swan" typhoon events that may occur only once

Q. Li et al., *A Hybrid Data-Model and AI-Driven Approach for Structural Monitoring in Hazardous Construction*, https://doi.org/10.1007/978-981-95-8688-2_3

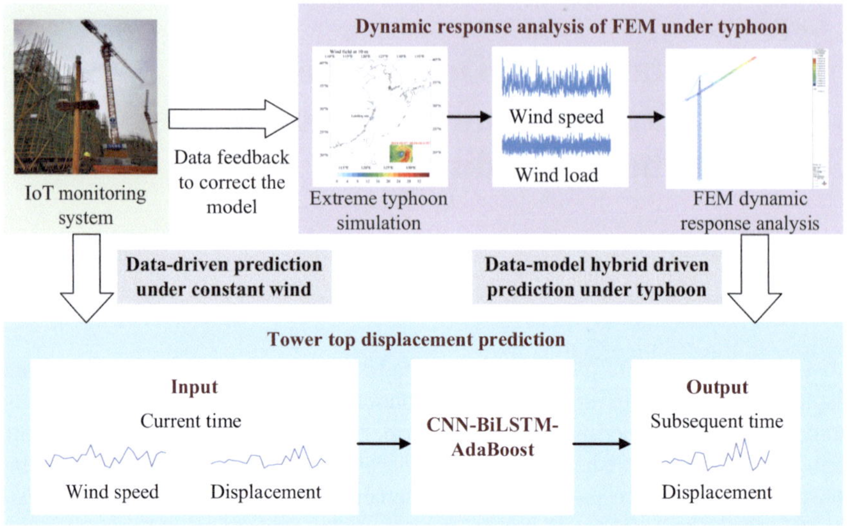

Fig. 3.1 Data model hybrid driven framework for tower top displacement response prediction

in a century or even once in a millennium, prediction approaches based solely on typical benign wind conditions struggle to accurately estimate structural responses and tend to produce underestimations. Therefore, it is necessary to improve conventional data-driven methods to enhance the accuracy of structural response prediction under such extreme conditions.

To address these challenges, this chapter proposes a hybrid data-model-driven framework for predicting tower-top displacement responses of tower cranes under extreme typhoon loading, as illustrated in Fig. 3.1. The framework consists of three core modules. The first module is an IoT-based monitoring system, which monitors on-site wind speed and tower crane displacement. The collected data are used to optimize and calibrate the parameters of the FEM in the second module and can also be applied to displacement prediction under benign wind conditions. The second module is a finite element dynamic response analysis module under typhoon loading, in which full-path typhoon simulation and wind field simulation methods are employed to compute the mean wind speed and fluctuating wind speed at the site. Subsequently, a FEM of the tower crane is established and calibrated using measured data. By applying wind load time histories to the calibrated model and performing multiple dynamic response analyses, a large typhoon wind speed-displacement dataset is generated. These datasets serve as a supplement to measured data and provide sufficient extreme-condition training samples for the third module. The third module focuses on tower-top displacement prediction, where the typhoon wind speed-displacement dataset is used for model training. The model takes the current wind speed and displacement as input variables and applies the CNN-BiLSTM-AdaBoost algorithm to predict the displacement at the next time step.

3.1 IoT-Based Real-Time Monitoring System for Tower Cranes

3.1.1 Monitoring Equipment

The No. 1 self-erecting tower crane at a construction site in Ningbo is shown in Fig. 3.2. The crane model is QTZ63. The tower height is 51 m, and the crane adopts integral standard sections with dimensions of 1.6 m × 1.6 m × 2.5 m. The jib length is 50 m, and the rated lifting moment is 630 kN·m.

To facilitate real-time observation of the operational status of the tower crane, WEMS401-LTL inclination sensors and WEMS401-VIB acceleration sensors were selected, both of which support real-time data transmission. The sensors were mounted at the measurement points using clamp fixtures, allowing for long-term monitoring under harsh environmental conditions, as shown in Fig. 3.3. To analyze the displacement behavior of different segments of the tower crane under typhoon loading, the instrumentation layout is illustrated in Fig. 3.4. A total of seven inclination sensors were installed: Sensors No. 1–6 were arranged along the standard sections of the tower mast, while Sensor No. 7 was fixed at the tower top. The inclination sensors measure tri-axial tilt angles along the X, Y, and Z axes, and the uploaded data includes the average, maximum, and minimum values of the tri-axial tilt angles. The measurement range is $\pm$ 90°, with a sensitivity of 0.001° and a maximum error of 0.01°. The acceleration sensor was installed at the No. 6 position to identify the dynamic characteristics of the crane. This sensor measures tri-axial accelerations along the X, Y, and Z axes, and the uploaded data includes the average, maximum, and minimum vibration values within each sampling period. The measurement range

Fig. 3.2 QTZ63 tower crane

is ± 2 g, with a sensitivity of 0.07 mg. To enable more intuitive observation of the operational status of the tower crane, the data acquisition frequency of the system was set to once per minute.

Since the research is closely related to wind speed, one wind speed monitoring device with data storage and real-time monitoring capabilities was installed at the site

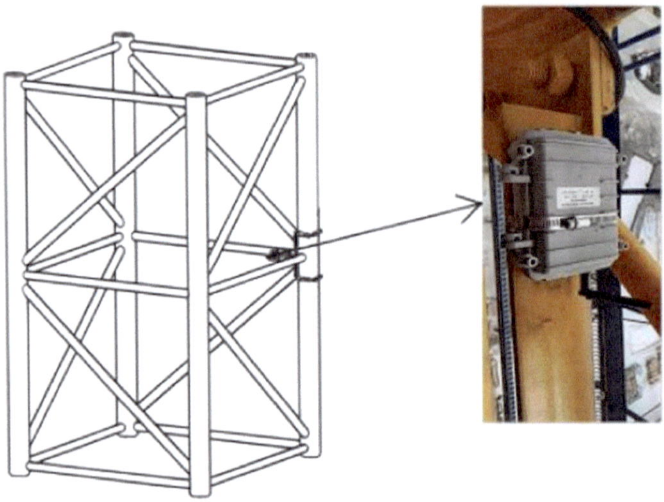

Fig. 3.3 Sensor arrangement positions on standard tower sections

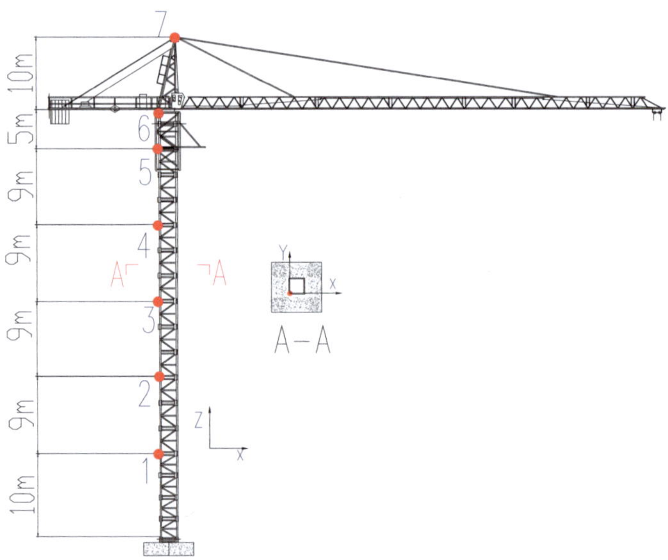

Fig. 3.4 Instrument layout and coordinate system

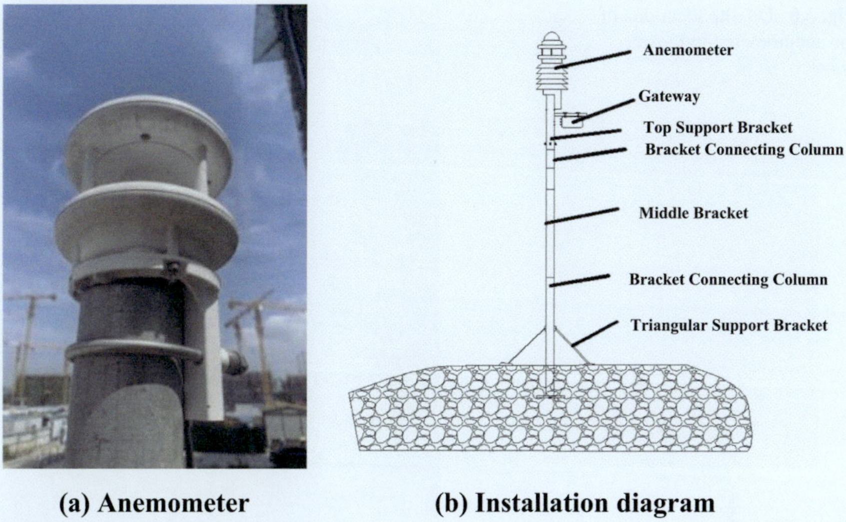

Fig. 3.5 Wind speed monitoring device

at a height of 10 m above ground to monitor variations in on-site wind conditions, with a sampling frequency of 4 Hz. The instrument can automatically and continuously record changes in wind speed and wind direction and transmit the data to the terminal via a gateway in real time. The uploaded data includes the current wind speed, minimum wind speed, maximum wind speed, and average wind speed. The wind speed monitoring device and its installation method are shown in Fig. 3.5, and the relative position between the wind speed monitoring device and the tower crane at the site is illustrated in Fig. 3.6.

3.1.2 Monitoring Cloud Platform

A complete wireless sensor network consists of a perception layer, a transmission layer, and an application layer. The perception layer is responsible for data acquisition, the transmission layer for data transfer, and the application layer for data processing and visualization. Figure 3.7 illustrates the topology of the tower crane IoT system, which in this study is likewise divided into three layers. The perception layer is composed of inclination sensors, acceleration sensors, and an anemometer, and is used to collect data such as the tower crane's inclination, acceleration, and on-site wind speed during operation; the network layer mainly relies on a GPRS network to enable wireless data transmission; and the application layer consists of a remote monitoring cloud platform, which is used to display real-time data, as shown in Fig. 3.8. To achieve intelligent management of the tower crane, users log into the system with a designated username and password to check the operating status of the

Fig. 3.6 On-site locations of the anemometer and tower crane

sensors based on the above hardware configuration. In addition, the cloud platform provides data storage and download functions, which not only ensure data security but also facilitate more convenient and efficient monitoring of the tower crane.

3.1.3 Monitoring Data Analysis

The variation of inclination is shown in Fig. 3.9. The original placement of the sensor forms a three-dimensional Cartesian coordinate system defined by the X, Y, and Z axes. When the tower mast undergoes lateral displacement, the sensor tilts accordingly, and its coordinate system changes to a new three-dimensional Cartesian system defined by the X', Y', and Z' axes. Here α denotes the angle between the X' axis and the horizontal plane, β denotes the angle between the Y' axis and the horizontal plane, and θ denotes the angle between the Z' axis and the horizontal plane. In Fig. 3.10, OA represents the height h of a specific segment of the tower mast. By converting the inclination angles measured by the sensors into the lateral displacements of each segment and the coordinates of the segment endpoints (x, y, z), using Eq. (3.1), the displacement of the tower top can be obtained.

$$\begin{cases} x = h \times \sin(\alpha) \\ y = h \times \sin(\beta) \\ z = h \times \sin(\theta) \end{cases} \tag{3.1}$$

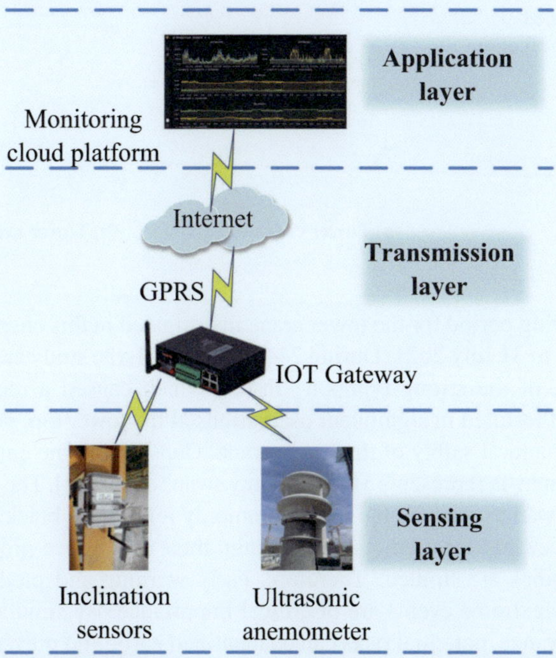

Fig. 3.7 Topology of the tower crane IoT system

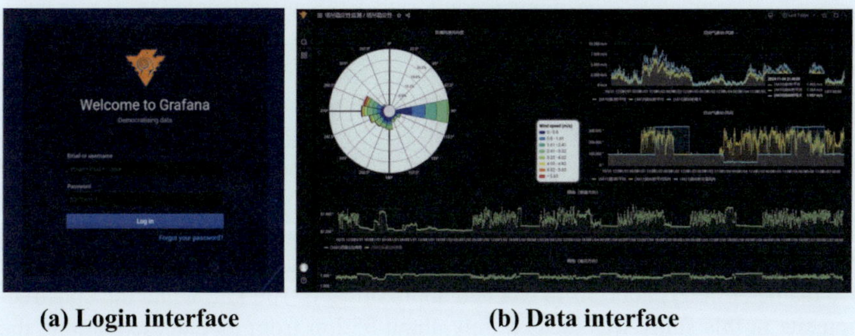

Fig. 3.8 Tower crane monitoring cloud platform

Fig. 3.9 Illustration of tilt angle variations

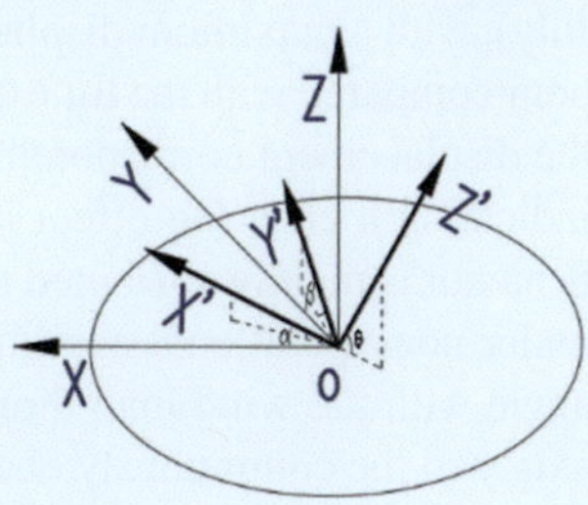

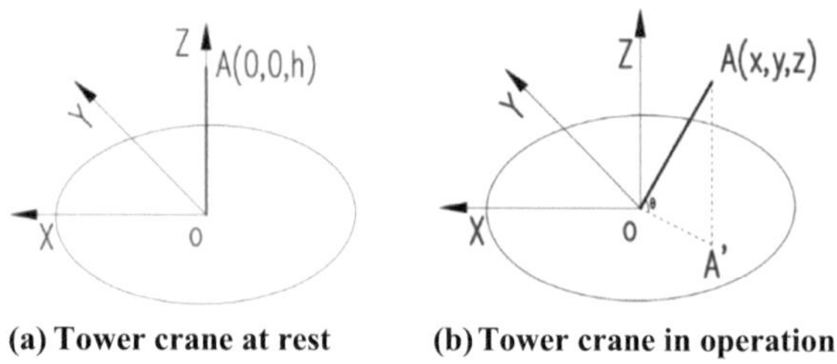

Fig. 3.10 Simplified computational model of the tower crane

The monitoring period for the tower crane investigated in this chapter spans from 1 January 2021 to 31 July 2021. During 24–26 July 2021, the study area was affected by the passage of the strong typhoon "In-fa", which caused a rapid increase in wind speed and resulted in significant oscillations at the tower top, posing a serious threat to the structural safety of the tower crane. Considering the entire monitoring period, Typhoon In-fa represents a typical "gray swan" event [64]. The key distinction between gray swan events and the more commonly referenced black swan events is that gray swan events are predictable, although their occurrence probability is low and related studies are limited. Therefore, early warning and prediction of such low-probability extreme events are of critical importance. By simulating these risk scenarios in advance, potential risks can be identified early, and preventive measures can be taken to mitigate their adverse impacts.

The raw monitoring data were downloaded from the monitoring cloud platform (Fig. 3.8). The uploaded one-minute monitoring data were resampled to 10-min intervals, and the corresponding maximum 3 s gust wind speed and tower-top displacement within each interval were extracted. Figure 3.11 shows a portion of the processed 3 s gust wind speed data and the corresponding tower-top displacement responses, including the typhoon period. As shown in Fig. 3.11a, the on-site wind speed began to increase from sample No. 3760 (14:00 on 24 July) and reached a maximum value of 40.86 m/s at sample No. 3866 (07:40 on 25 July). Notably, at sample No. 3952 (22:00 on 25 July), the wind speed dropped to a minimum of 4.59 m/s, resulting in a maximum daily wind speed difference of 36.27 m/s. According to related reports, the eye diameter of Typhoon In-fa reached approximately 100 km, while the construction site in this study was located only about 35 km from its path, explaining the pronounced wind speed variation within the same day [61]. As shown in Fig. 3.11b, the maximum tower-top displacement occurred at sample No. 3874 (09:00 on 25 July), with a maximum displacement of 648.73 mm, which lagged by more than one hour compared with the time of maximum wind speed (07:40 on 25 July). In contrast, the displacement corresponding to the maximum wind speed was only 200.44 mm, indicating a clear lag effect in the displacement response of the tower crane. This behavior is mainly attributed to the wind-induced rotational effect of the tower crane under non-operational conditions. When the upper structure of the crane forms an angle with the wind load direction, the slewing mechanism rotates about the crane axis, and the continuously changing windward area leads to delayed load responses. When the crane aligns parallel to the wind direction and rotation ceases, the windward area becomes minimal, and the displacement gradually decreases.

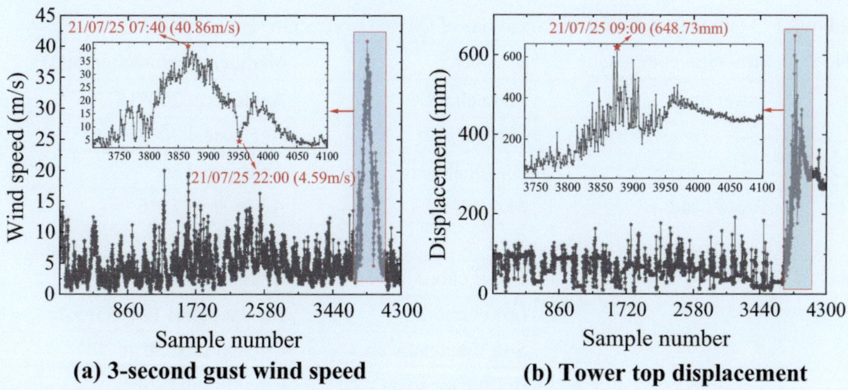

Fig. 3.11 Wind speed and corresponding tower top displacement

3.2 Finite Element Dynamic Response Analysis of Tower Cranes Under Extreme Typhoon Loading

3.2.1 Development and Calibration of the Finite Element Model

The QTZ63 tower crane is a multipurpose tower crane with a rated lifting moment of 630 kN·m, a maximum lifting capacity of 6 t, and a maximum working radius of 55–56 m, and is one of the most commonly used medium-sized tower cranes on construction sites in China. In this section, this type of tower crane is taken as the research object, and a FEM is established using the large-scale general-purpose finite element software Midas Gen. The specific parameters are listed in Table 3.1. All structural components are made of Q235 steel, with an elastic modulus E = 210 GPa, Poisson's ratio $\mu = 0.3$, and density $\rho = 7.85$ kg/m^3.

To ensure the rationality of the model, the FEM of the tower crane should be established by comprehensively considering the following aspects: (1) the model should be able to fully and accurately represent the structural characteristics of the tower crane; (2) the loading conditions of the model should be consistent with the actual operating conditions of the tower crane; and (3) the boundary conditions of the model should conform to the real working conditions of the tower crane. Based on these principles, the tower crane model is simplified. The main components considered include the tower mast, tower top, jib, counter-jib, and slewing mechanism. Rigid connections are assumed between the segments of the tower mast and jib, and shared nodes are adopted. The crossbeams and columns of the tower mast and jib, as well as the slewing mechanism, are equivalently modeled using beam elements, while diagonal bracings are modeled using truss elements to reduce the number of element types and facilitate computation. The effects of the trolley, hook, tower mast ladder, and electrical equipment on the dynamic characteristics of the tower crane

Table 3.1 Main materials and specifications of QTZ63 tower crane components

No	Structural component	Member	Member specification (mm)
1	Tower mast	Main chord	Angle steel 200*16
		Vertical web	Angle steel 70*7
		Diagonal web	Angle steel 86*7
2	Tower head	Main chord	Angle steel 75*6
		Web	Circular tube 76*5
3	Jib	Upper chord	Circular tube 89*8
		Lower chord	Channel steel 120*53*5.5
		Side diagonal web	Circular tube 50*4
		Bottom diagonal web	Circular tube 30*3
4	Counter-jib	Main chord	No. 14 I-beam
5	Counter-jib tie rod		Circular steel tube, d = 56
6	Jib tie rod		Circular steel tube, d = 60

are neglected. However, neglecting these components may lead to differences in mass distribution between the model and the actual structure, thereby affecting the accuracy of the dynamic response of the tower crane under wind loading. Therefore, the density of the tower crane structure is adjusted so that the model has a mass distribution similar to that of the actual structure.

In simulation analysis, the rationality of boundary condition settings directly determines the accuracy of the results. The base of the tower mast has relatively high stiffness and is therefore modeled as a fixed support. The connections at the root of the jib, the connections between the jib and the two tie rods, the connections between the counter-jib and the two tie rods, and the connections between the tie rods and the tower head are all modeled as pinned supports. The slewing bearing connections to both the tower head and the tower mast are modeled as fixed connections. The final FEM of the tower crane is shown in Fig. 3.12, where the X-axis denotes the jib extension direction, the Y-axis denotes the direction perpendicular to the tower mast, and the Z-axis denotes the tower mast extension direction.

Finite element analysis of tower crane responses often exhibits discrepancies when compared with actual monitoring results. Therefore, it is necessary to periodically calibrate the FEM using measured response data to ensure consistency between the model and the actual structure and to obtain accurate response predictions. Based on engineering experience, several structural parameters that have a significant influence on tower crane responses are selected for sensitivity analysis. Among them, the dimensions of the tower mast main chord angle steel, the wall thickness of the upper chord tube of the jib, the section size of the lower chord channel steel of the jib, and the diameter of the tie rods have a strong influence on the displacement response of the tower crane and are therefore identified as calibration parameters. The sensitivity calculation formula is given as follows:

Fig. 3.12 Schematic Diagram of the Tower Crane FEM

$$\phi_i = \frac{\Delta Y/Y}{\Delta\theta_i/\theta_i} \tag{3.2}$$

In the equation, ϕ_i denotes the parameter sensitivity, Y represents the finite element calculation result, i.e., the structural displacement response, and θ_i is the calibration parameter.

For tower crane systems, extracting only the lower-order modes is sufficient to adequately characterize their dynamic properties. Accordingly, the first three natural frequencies of the tower crane are selected as the structural characteristic quantities for FEM updating in this study. Modal analysis is performed again on the updated FEM to obtain the calibrated dynamic characteristics of the tower crane. The measured modal frequencies of the tower crane and those before and after model calibration are listed in Table 3.2. The results indicate that the updated FEM shows good agreement with the measured values for the first three natural frequencies, with errors reduced to within 1%, and the RMSE significantly decreased from 0.0456 Hz to 0.0075 Hz, demonstrating that the calibrated model provides a much closer match to the measured frequencies. Figure 3.13 presents the first three mode shapes of the calibrated tower crane model. The first mode corresponds to rotation of the jib about the tower mast in the x–y plane; the second mode represents fore-aft sway of the tower mast accompanied by bending deformation of the jib; and the third mode corresponds to slight tilting of the tower mast with torsional deformation of the jib. It can be observed that the vibration modes of the tower crane are dominated primarily by deformation of the jib, followed by deformation of the tower mast, which is consistent with the actual dynamic behavior of tower cranes.

Table 3.2 First three natural frequencies of vibration modes

Mode	Measured value (Hz)	FEM before calibration (Hz)	Error before calibration (%)	FEM after calibration (Hz)	Error after calibration (%)
1	0.4107	0.4322	5.24	0.4141	0.83
2	0.8975	0.9110	1.49	0.9005	0.33
3	1.5019	1.5769	5.05	1.5128	0.70

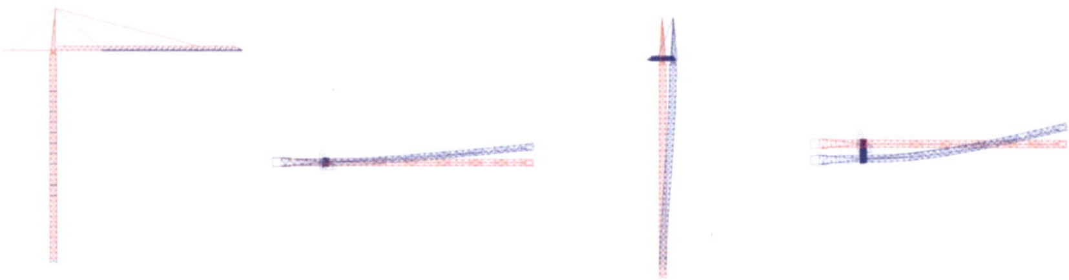

(a) First-order mode shape (f_1 = 0.4141 Hz) **(b) Second-order mode shape (f_2 = 0.9005 Hz)**

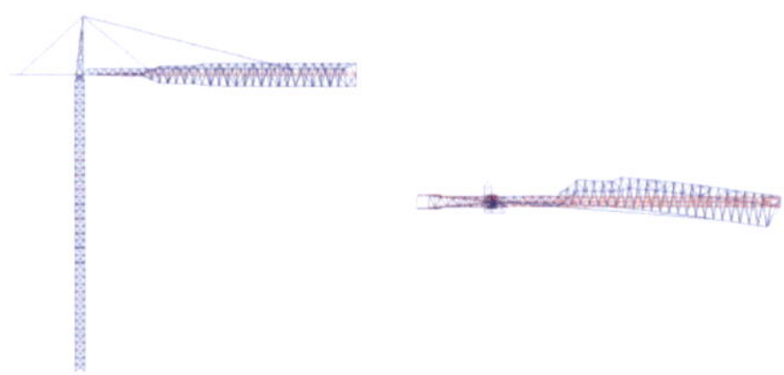

(c) Third-order mode shape (f_3 = 1.5128 Hz)

Fig. 3.13 First three mode shapes of the calibrated FEM

3.2.2 *Simulation of Extreme Typhoon Conditions*

Wind climate assessment can determine the extreme wind speeds from different directions at a construction site, providing a basis for the mean wind speed input used in the dynamic analysis of tower cranes. The Ningbo region belongs to a mixed wind climate zone and is influenced throughout the year by both the East Asian monsoon and typhoons originating from the Northwest Pacific.

Due to the limited number of historical typhoon wind speed observations, this study adopts a full-track typhoon simulation approach. By applying the Yan Meng wind field model, near-surface typhoon wind speeds affecting the Ningbo region over a 600-year period are obtained [97]. A single full-track typhoon simulation can be divided into four stages, as illustrated in Fig. 3.14. In the first stage, the typhoon genesis location is randomly selected based on historical typhoon occurrence data. In the second stage, a movement model and an intensity model are established and updated every 6 h to simulate the typhoon's motion and evolution. The third stage employs a filling-rate model to simulate the typhoon landfall process. In the final stage, the Yan Meng wind field model is activated to simulate the region within a 250 km radius around Ningbo. The Vickery empirical model [98] is adopted to determine key wind field parameters, including the radius of maximum wind speed *Rmax* and the Holland pressure profile parameter B, with an equivalent surface roughness length of 0.05 m. To ensure a sufficient number of annual extreme values for statistical analysis, a 600-year Monte Carlo simulation is conducted to estimate the typhoon climate of the Ningbo region, and the detailed procedures can be found in Refs. [97, 99, 100]. Using the simulated typhoon wind speeds together with the steady wind speeds recorded at meteorological stations, joint probability distribution models of annual maximum wind speeds for synoptic winds and typhoons under 16 wind directions in the Ningbo region are constructed based on the t-Copula function. The probability distributions of annual maximum wind speeds for each wind direction are shown in Fig. 3.15a, and the extreme wind speeds corresponding to a 100-year return period are presented in Fig. 3.15b. The results indicate that the extreme wind speeds induced by typhoons exceed those of synoptic winds in all directions, demonstrating that the wind climate in Ningbo is dominated by typhoons. Furthermore, analysis of the extreme wind speeds from different directions shows that D2 is the most unfavorable direction, with an extreme wind speed of 41.4 m/s. Therefore, in the subsequent wind-induced dynamic response analysis of the tower crane, the upper limit of the simulated wind speed is set to 41.4 m/s.

3.2.3 Dynamic Response Analysis of the Tower Crane Using FEM

Figure 3.16 illustrates the segmentation of the tower crane and the locations of the simulation points. The tower crane is divided into 18 segments, with the bolted joints defined as the load application points, and the simulated wind loads are applied to the corresponding nodes along the vertical direction in the direction of the jib (i.e., the Y-axis). Through dynamic response analysis, the displacement response at the tower top can be obtained. It should be noted that under extreme typhoon conditions, the tower crane is in a non-operational state. According to the Chinese crane design code GB3811-2008, the wind load acting on a non-operational tower crane should consider the wind pressure height variation coefficient; therefore, the variation of wind speed with height and the characteristics of fluctuating wind are taken into account in the simulation of wind speed time histories. In this study, the harmonic superposition method is employed to simulate the wind speed, and based on the wind climate assessment results, 41.4 m/s is adopted as the upper limit of the basic wind speed. The ground roughness is classified as Category B, with the ground roughness coefficient set to 0.16, and the Kaimal spectrum is used to represent the wind spectrum. Subsequently, the wind load time histories for each simulation region are calculated. Figure 3.17 presents the wind load time histories at several key nodes

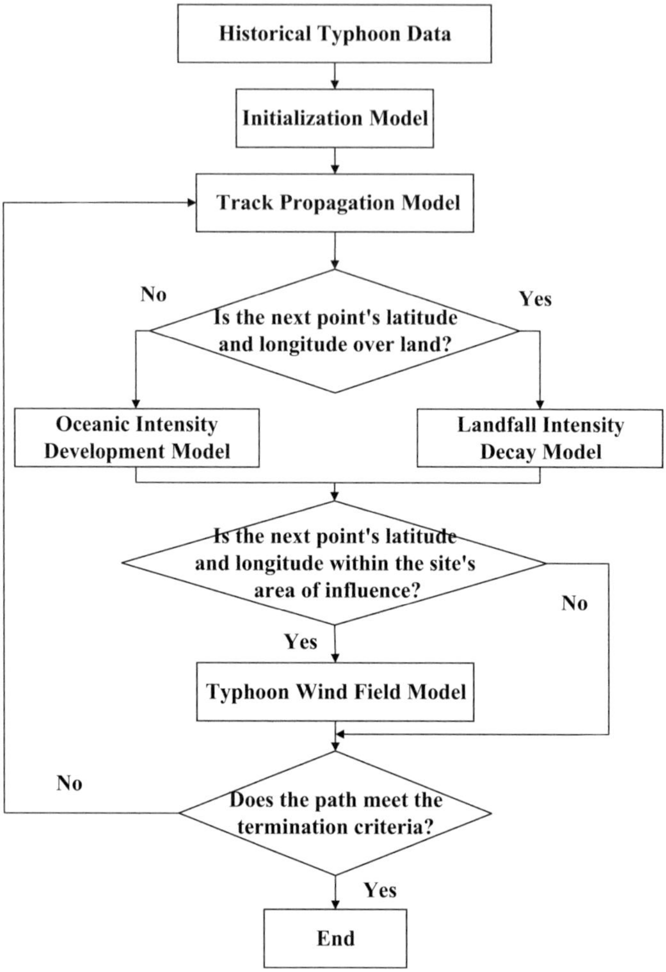

Fig. 3.14 Flowchart of the full-track simulation process for a single typhoon

of the tower crane when the basic wind speed is 41.4 m/s, from which it can be observed that the wind load increases with height.

Considering that the tower crane does not operate during typhoon events, the influence of lifting loads on the response results must be eliminated during model validation. Figure 3.18 presents the measured data under non-operational conditions (i.e., zero lifting load) within the two weeks prior to the arrival of Typhoon "In-fa", together with their exponential fitting curves. For comparison, the simulated wind speeds and the corresponding displacement responses of the calibrated FEM are also fitted. It can be observed from the figure that the trends of the fitted curves are consistent. In addition, the figure indicates the displacement values corresponding to the

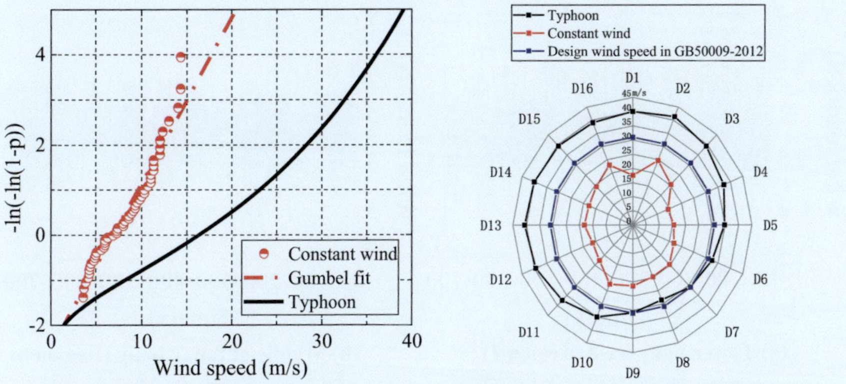

(a) Probability of annual maximum wind speed by D2 wind direction

(b) Extreme wind speeds for different wind directions under a 100-year return period

Fig. 3.15 Wind climate assessment of the Ningbo Region

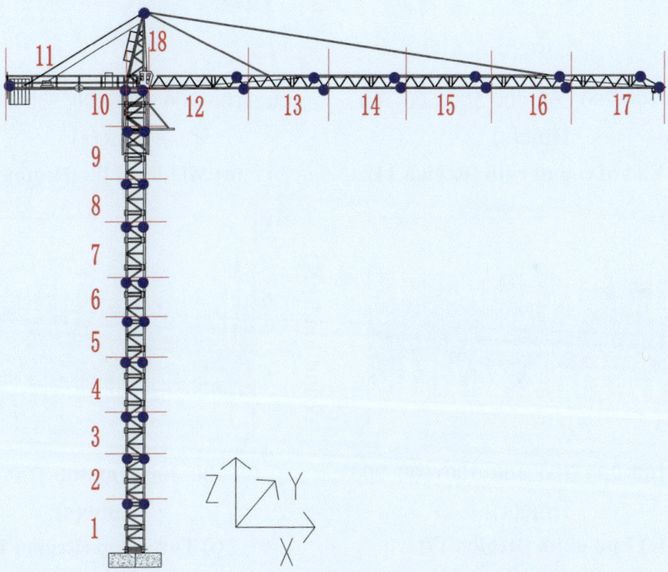

Fig. 3.16 Schematic diagram of tower crane segments and simulation points

two fitted curves at wind speeds of 5, 10, and 15 m/s, where the maximum displacement difference is 6.03 mm, demonstrating that the updated model can accurately reflect the actual structural loading and deformation behavior.

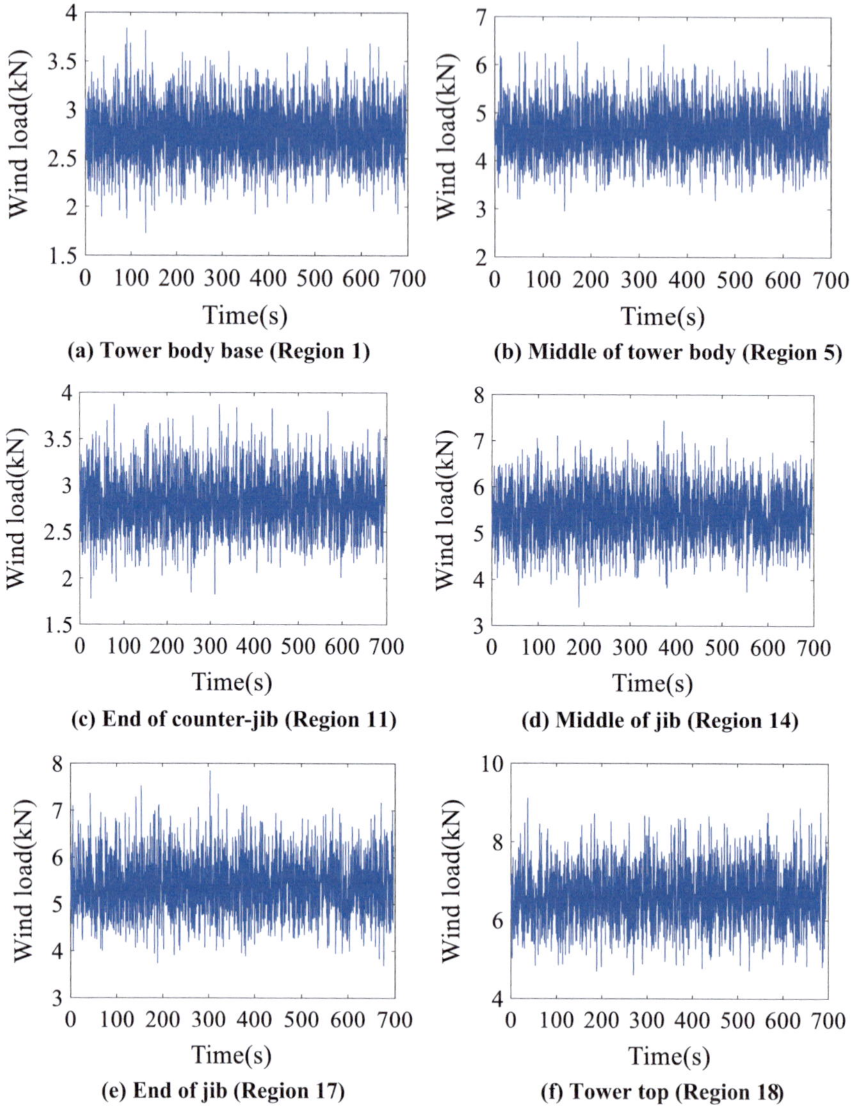

Fig. 3.17 Wind load time history with a basic wind speed of 41.4 m/s

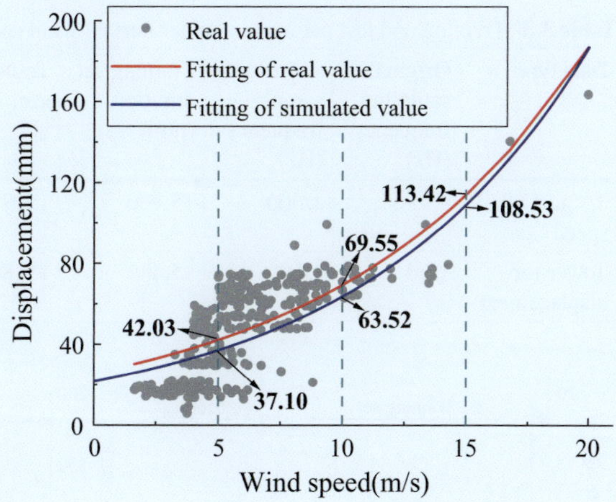

Fig. 3.18 Comparison of model calibration results

3.3 Machine Learning-Based Prediction of Tower Crane Dynamic Responses

3.3.1 Data-Driven Prediction of Tower-Top Displacement

In this chapter, the data-driven prediction of tower-top displacement uses the maximum s gust wind speed and the corresponding tower-top displacement at the current time step as inputs, while the maximum tower-top displacement at a subsequent time step is taken as the output to train the prediction models. All prediction models adopt the CNN-BiLSTM-AdaBoost algorithm. The convolutional layer contains 16 convolution kernels with a kernel size of 3×1; the pooling layer employs max pooling with a window size of 2×1; the BiLSTM layer consists of two single LSTM layers, each with 25 units; and the AdaBoost layer constructs a total of three weak learners. During prediction, changes in the daily operational procedures of the tower crane at the construction site must be considered. If the prediction lead time is set too long (e.g., half a day or longer) or too short, resulting in insufficient emergency response time, the value of early warning will be reduced. After in-depth communication and discussion with the project staff, the prediction lead time for the tower crane response is determined to be 10 min in this study.

(1) **Prediction of tower-top displacement response under Normal Wind Conditions**

To conduct the prediction of the tower crane response under normal wind conditions, part of the data under normal wind conditions shown in Fig. 3.11 (a total of 2150 data sets) is divided into two subsets, as presented in Table 3.3 and Fig. 3.19. The first subset is the training set, consisting of 1548 samples and accounting for 72% of the dataset, while the second subset is the testing set, consisting of 602 samples and accounting for 28% of the dataset.

Table 3.3 Training and test parameters under normal wind conditions

Data type	Original sampling frequency (Hz)	Adjusted sampling frequency (Hz)	Training set duration (min)	Training set samples	Testing set duration (min)	Testing set samples
3-s gust wind speed	4	1/600	15,480	1548	6020	602
Tower-top displacement	1/60	1/600	15,480	1548	6020	602

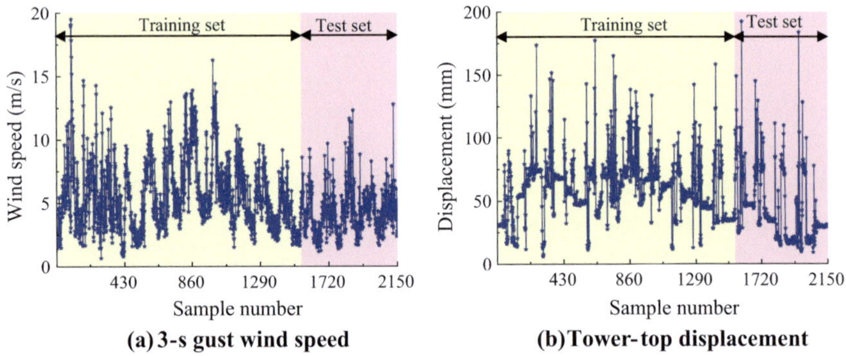

Fig. 3.19 Dataset partitioning under normal wind conditions

Figure 3.20 presents the prediction results of the horizontal displacement response at the tower top under normal wind conditions. As shown in Fig. 3.20a, the predicted values obtained by the CNN-BiLSTM-AdaBoost, CNN, and LSTM models are compared with the measured values. It can be observed that the predictions of the CNN-BiLSTM-AdaBoost model are closer to the measured values, demonstrating stronger short-term prediction capability and better robustness, whereas the CNN and LSTM models exhibit certain deviations from the actual values. Figure 3.20b shows the prediction errors of different models, from which it can be seen that the CNN-BiLSTM-AdaBoost model has relatively smaller and more stable errors. The regression plots of the CNN-BiLSTM-AdaBoost model shown in Fig. 3.20c–d indicate a strong correlation between the measured data and the predicted values, with an R^2 value of 0.9932 for the training set and 0.9910 for the testing set. As the slope of the regression curve is close to 1, the model exhibits good prediction performance on both the training and testing sets. To further compare the prediction accuracy of different models, the error metrics of each model are calculated, as listed in Table 3.4. As shown in Table 3.4, the CNN-BiLSTM-AdaBoost model achieves the smallest RMSE and overall error, and its R^2 value is closest to 1, indicating the best prediction performance. To more intuitively analyze the prediction errors of extreme samples, four samples with the largest measured displacements (sample numbers 44, 413,

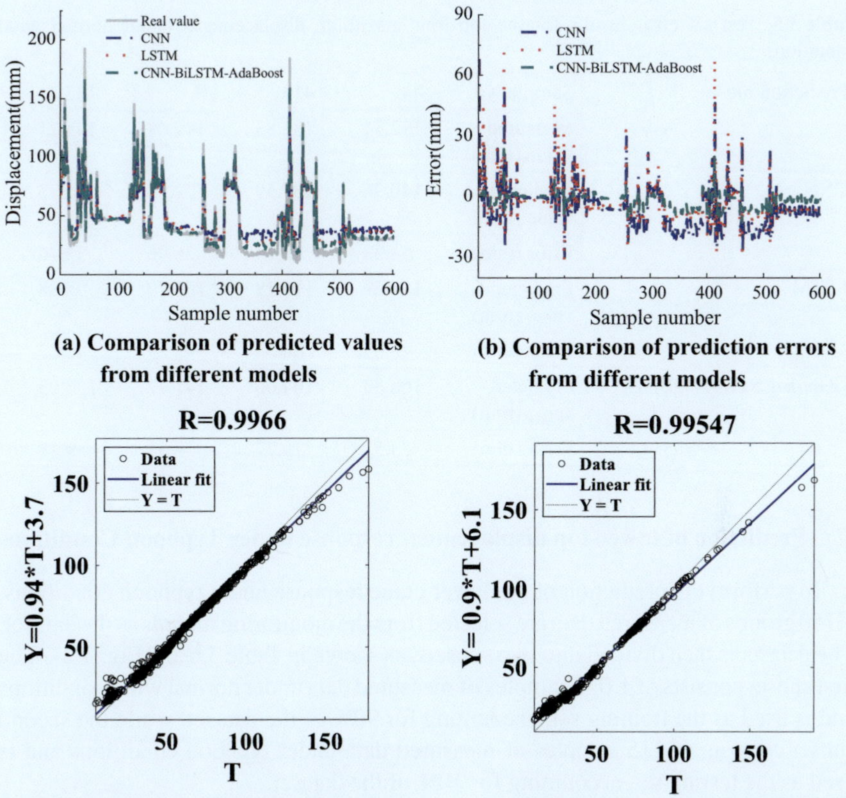

(a) Comparison of predicted values from different models

(b) Comparison of prediction errors from different models

(c) Regression curve for the training set

(d) Regression curve for the test set

Fig. 3.20 Prediction results of tower-top horizontal displacement under normal wind conditions

Table 3.4 Prediction metrics of different models under normal wind conditions

Prediction model	RMSE (mm)	R^2	Maximum error (mm)
CNN	11.3171	0.9343	51.38
LSTM	10.2542	0.9399	75.05
CNN-BiLSTM-AdaBoost	3.8329	0.9910	23.51

8, and 132) are selected from the testing set, and the corresponding measured and predicted displacements are listed in Table 3.5. The CNN-BiLSTM-AdaBoost model still demonstrates superior performance, with a maximum prediction error of 23.51 mm, which corresponds to a tower crane inclination rate error of 3.9/10,000 and can be considered negligible. Overall, the CNN-BiLSTM-AdaBoost model exhibits good fitting capability and high prediction accuracy for the tower-top displacement under normal wind conditions.

Table 3.5 Test set error results for the top four maximum displacements under normal wind conditions

Prediction model	Sample No.	44	413	8	132
	Measured value (mm)	192.34	183.83	149.08	145.21
CNN	Predicted value (mm)	140.96	137.59	122.12	111.75
	Error (mm)	51.38	46.24	26.96	33.46
LSTM	Predicted value (mm)	117.29	115.28	105.79	98.88
	Error (mm)	75.05	68.55	43.29	46.33
CNN-BiLSTM-AdaBoost	Predicted value (mm)	168.83	164.06	141.77	136.03
	Error (mm)	23.51	19.77	7.31	9.18

(2) **Prediction of tower-top displacement response under Typhoon Conditions**

To perform the prediction of the tower crane response under typhoon conditions, 1,150 groups of measured data are selected from the monitoring records as the dataset. The dataset is then divided into two subsets, as shown in Table 3.6 and Fig. 3.21. The first subset consists of 1,035 samples of measured data under normal wind conditions and is used as the training set, accounting for 90% of the dataset, while the second subset consists of 115 samples of measured data under typhoon conditions and is used as the testing set, accounting for 10% of the dataset.

Figure 3.22 presents the prediction results of the horizontal displacement response at the tower top under typhoon conditions. As shown in Fig. 3.22a, the predicted values of the three models are compared with the measured values, and it cn be observed that all models exhibit certain deviations from the actual values. Figure 3.22b shows the prediction errors of different models, from which it can be seen that the LSTM model has large error fluctuations, whereas the CNN-BiLSTM-AdaBoost model exhibits relatively smaller and more stable errors. From the regression plots of the CNN-BiLSTM-AdaBoost model shown in Fig. 3.22c–d, although the testing data show a strong correlation, the slope of the regression curve deviates

Table 3.6 Training and test parameters under typhoon conditions

Data type	Original sampling frequency (Hz)	Adjusted sampling frequency (Hz)	Training set duration (min)	Training set samples	Testing set duration (min)	Testing set samples
3-s gust wind speed	4	1/600	10,350	1035	1150	115
Tower-top displacement	1/60	1/600	10,350	1035	1150	115

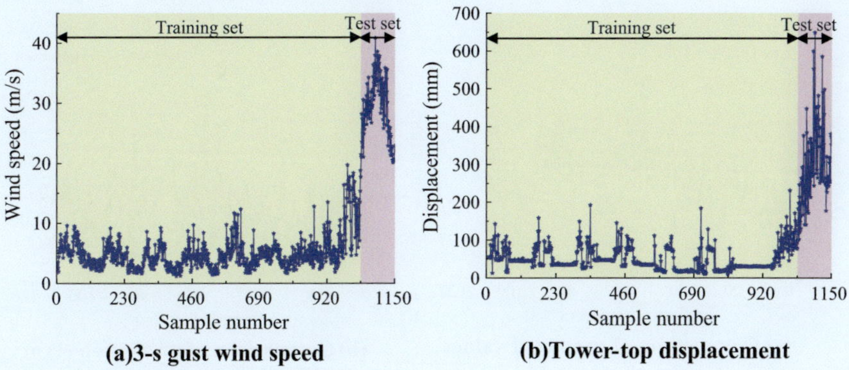

Fig. 3.21 Dataset partitioning under typhoon conditions

significantly from 1, indicating relatively large prediction errors, which increase with the magnitude of displacement. Table 3.7 further quantifies the results shown in the figure: the CNN model has relatively small RMSE and errors but a lower R^2 value, indicating poor fitting performance; the prediction accuracy of the CNN-BiLSTM-AdaBoost model is better than that of the LSTM model but slightly lower than that of the CNN model, with an R^2 value of 0.9594, indicating relatively good fitting performance. Therefore, overall, the CNN-BiLSTM-AdaBoost model performs best among the three models. In addition, four samples with the largest measured displacements (sample IDs 57, 52, 82, and 90) are listed, and their prediction errors are shown in Table 3.8. It is worth noting that the prediction errors for these extreme samples are very large, with the maximum error of the CNN-BiLSTM-AdaBoost model reaching 247.41 mm, indicating that the tower crane displacement prediction model significantly underestimates the response during "grey swan" typhoon events. In summary, using a purely data-driven approach to predict the structural displacement response of a tower crane under extreme typhoon conditions has inherent limitations. Due to the lack of response data under extreme typhoon events in the training dataset, the prediction models have insufficient learning of extreme conditions and tend to underestimate the response when exposed to typhoon conditions, thereby failing to provide accurate displacement predictions.

3.3.2 Hybrid Data-Model-Driven Prediction of Tower-Top Displacement

In this section, the prediction of the tower crane response under typhoon conditions is further investigated, and a dataset consisting of 1,150 samples is constructed. The simulated extreme typhoon wind speeds obtained in Sect. 3.2 and the corresponding extreme displacement data derived from finite element simulations are used as the training set, comprising 1,035 samples (accounting for 90% of the dataset), while the

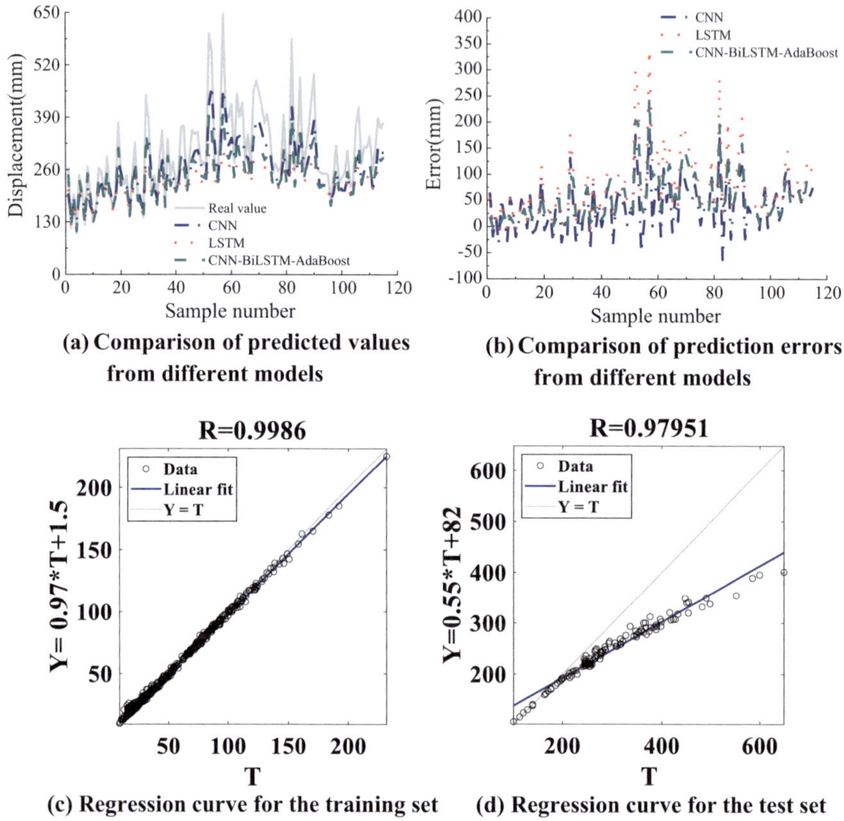

Fig. 3.22 Prediction results of tower-top horizontal displacement under typhoon conditions

Table 3.7 Comparison of prediction metrics of different models under typhoon conditions

Prediction model	RMSE (mm)	R^2	Maximum error (mm)
CNN	62.2678	0.8695	191.09
LSTM	115.5612	0.8771	340.65
CNN-BiLSTM-AdaBoost	71.2089	0.9594	247.42

testing set consists of 115 samples (accounting for 10%) from the actually monitored data under typhoon conditions. The testing set is identical to that used in Sect. 3.3.1 for the data-driven prediction of tower-top displacement under typhoon conditions. The sampling frequency, duration, and sample size are detailed in Table 3.6, and the dataset partition is illustrated in Fig. 3.23.

Figure 3.24 presents the prediction results of the horizontal displacement response at the tower top under typhoon conditions. As shown in Fig. 3.24a, the predicted values of the three models are compared with the measured values. It can be

Table 3.8 Test set error results for the top four maximum displacements under typhoon conditions

Prediction model	Sample No	57	52	82	90
	Measured value (mm)	648.48	599.51	585.19	497.68
CNN	Predicted value (mm)	457.39	423.48	415.34	381.02
	Error (mm)	191.09	176.03	169.85	116.66
LSTM	Predicted value (mm)	307.84	301.07	299.66	289.81
	Error (mm)	340.64	298.44	285.53	207.87
CNN-BiLSTM-AdaBoost	Predicted value (mm)	401.07	395.59	389.34	338.22
	Error (mm)	247.41	203.92	195.85	159.46

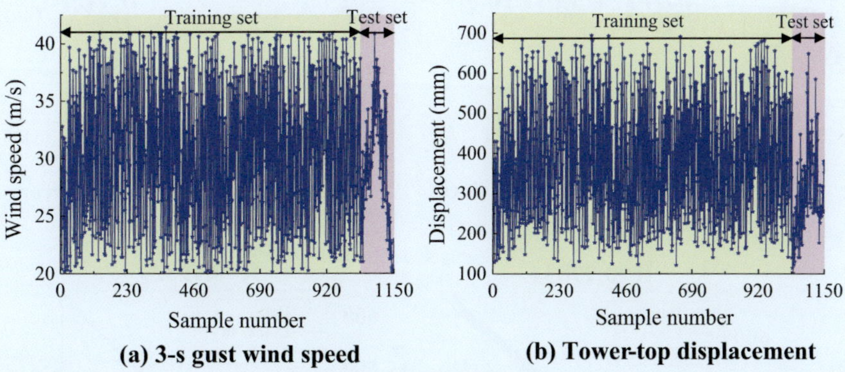

Fig. 3.23 Dataset partitioning under typhoon conditions (Hybrid Data-Model-Driven)

observed that the predictions of the CNN-BiLSTM-AdaBoost model are closest to the measured values, whereas the CNN and LSTM models exhibit certain deviations from the actual values. Figure 3.24b illustrates the prediction errors of different models, showing that the CNN-BiLSTM-AdaBoost model has relatively smaller and more stable errors, while the CNN and LSTM models display larger error fluctuations. From the regression plots of the CNN-BiLSTM-AdaBoost model shown in Fig. 3.24c–d a strong correlation between the measured and predicted values can be observed. Table 3.9 further quantifies the results presented in the figure, indicating that the CNN-BiLSTM-AdaBoost model achieves the best prediction performance under typhoon conditions, with a testing-set RMSE of 10.1283 mm, an R^2 value of 0.9906, and a maximum error of 33.53 mm. Compared with the purely data-driven response prediction method presented in Sect. 3.3.1, the prediction accuracy is significantly improved. Using the same evaluation procedure, the prediction errors for the four largest measured displacements in the testing set (sample IDs 57, 52, 82, and 90) are compared, as shown in Table 3.10. The results show that, for the CNN-BiLSTM-AdaBoost model, the maximum errors of the four samples are controlled at approximately 10 mm, indicating that the model can maintain small prediction errors even for samples with large measured displacements. In summary, the hybrid data-model-driven approach for predicting the tower-top displacement response under typhoon conditions effectively addresses the issue of insufficient extreme samples in the training dataset and improves the accuracy and reliability of predictions for "grey swan" events.

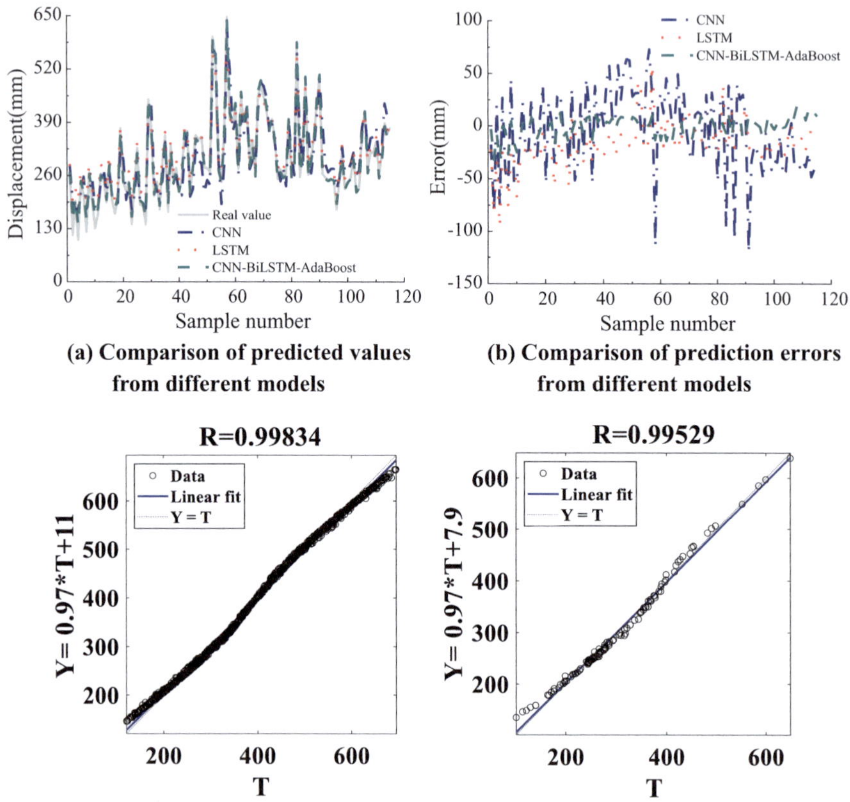

(a) Comparison of predicted values from different models (b) Comparison of prediction errors from different models

(c) Regression curve for the training set (d) Regression curve for the test set

Fig. 3.24 Prediction results of tower-top horizontal displacement under typhoon conditions (Hybrid Data-Model-Driven)

Table 3.9 Comparison of prediction metrics of different models under typhoon conditions (Hybrid Data-Model-Driven)

Prediction model	RMSE (mm)	R^2	Maximum error (mm)
CNN	39.8309	0.8526	119.95
LSTM	27.0768	0.9968	96.01
CNN-BiLSTM-AdaBoost	10.1283	0.9906	33.53

Table 3.10 Test set error results for the top four maximum displacements under typhoon conditions (Hybrid Data-Model-Driven)

Prediction model	Sample No	57	52	82	90
	Measured value (mm)	648.48	599.51	585.19	497.68
CNN	Predicted value (mm)	619.58	567.92	557.18	492.90
	Error (mm)	28.90	31.59	28.01	4.78
LSTM	Predicted value (mm)	589.58	555.61	545.16	478.25
	Error (mm)	58.90	43.90	40.03	19.43
CNN-BiLSTM-AdaBoost	Predicted value (mm)	638.51	597.26	585.30	506.47
	Error (mm)	9.97	2.25	0.11	8.79

3.4 Summary

This chapter proposes a hybrid data-model-driven approach for predicting the displacement response of a tower crane under extreme typhoon conditions. Using a tower crane in a real engineering project as a case study, an IoT-based real-time displacement monitoring system is developed to obtain load-response data for the tower crane during construction. In addition, a FEM is established to simulate the tower crane displacement under various extreme wind loads, thereby compensating for the lack of extreme samples in the monitored data. Based on the CNN-BiLSTM-AdaBoost algorithm, both data-driven and hybrid data-model-driven approaches are employed to predict the tower-top displacement under normal wind and typhoon conditions. According to the research results, the main conclusions are summarized as follows:

(1) Limitations of the purely data-driven approach

The purely data-driven approach exhibits significant limitations in predicting the displacement of a tower crane under extreme typhoon conditions and tends to underestimate the structural response of the tower crane.

(2) Improvement in prediction accuracy and robustness using the hybrid data-model-driven approach

By integrating real-time monitoring data with extreme typhoon simulation results, the hybrid data-model-driven approach significantly improves the prediction accuracy and robustness of tower crane displacement under extreme typhoon conditions.

Chapter 4
Real-Time Displacement Monitoring of High Formwork Support Structures Based on Computer Vision

Abstract The influence of various external factors on the proposed M-DAVIM visual measurement method is systematically investigated through laboratory tests and field monitoring. After identifying common environmental and operational influencing factors, a series of periodic vibration experiments is conducted under controlled conditions to analyze the effects of illumination intensity, fog concentration, camera elevation angle, target size, and camera vibration. A laser displacement sensor is employed as a reference to quantitatively evaluate measurement errors induced by these factors. Furthermore, the visual monitoring system is deployed at a high formwork construction site in Ningbo, enabling real-time tracking and displacement monitoring of critical structural nodes. The results demonstrate that under optimal lighting conditions (200–400 lx) and with target marker sizes exceeding 18 pixels, the M-DAVIM method achieves sub-millimeter measurement accuracy and maintains high precision and strong robustness under adverse conditions such as fog and vibration. Field monitoring results from May 27, 2023, show that all monitored indicators remain below the code-specified alarm thresholds, indicating structural safety. Nevertheless, due to continuous changes in structural conditions caused by factors such as component aging and load variation, there remains a need for scientific and adaptive methods to determine and dynamically update monitoring and warning thresholds.

Keywords High-formwork support system · Computer vision · M-DAVIM · External influencing factors

With the rapid development of the construction industry in China, building structures are becoming increasingly complex, taller, and characterized by larger spans. High formwork support structures are therefore widely used in engineering practice. Owing to their large height and small cross-sectional dimensions, the vertical posts of high formwork systems have large slenderness ratios. During concrete pouring or for a period after pouring, the applied loads may increase suddenly, which can easily lead to settlement and lateral deformation. As a result, the structure may lose overall stability and pose a threat to the safety of the entire construction project. To prevent

Q. Li et al., *A Hybrid Data-Model and AI-Driven Approach for Structural Monitoring in Hazardous Construction*, https://doi.org/10.1007/978-981-95-8688-2_4

accidents, it is necessary to monitor the settlement and deformation of the support system. Traditional displacement monitoring methods generally use instruments such as levels, theodolites, and total stations to measure structural displacement. These measurements require manual operation and can only be performed at one point at a time, resulting in disadvantages such as inconvenient operation, low efficiency, and the inability to conduct long-term continuous monitoring. In addition, although traditional displacement sensors can be installed near the support system for automatic measurement, their close proximity to the structure makes them susceptible to construction-induced vibrations, which adversely affect measurement accuracy.

Taking a high formwork support structure from a construction project in Ningbo as the case study, this chapter proposes a computer vision-based displacement measurement method to improve measurement accuracy and to achieve real-time construction monitoring and safety early warning. First, indoor experiments are conducted to investigate the influence of different external factors-such as illumination, target size, camera pose, and camera vibration-on the measurement accuracy of computer vision-based displacement monitoring. Subsequently, a real-time monitoring system for high formwork support structures based on computer vision is established at the construction site. By acquiring video data of the target structure during construction and combining it with a self-developed multi-point real-time displacement monitoring system, the structural displacement is obtained, enabling real-time monitoring and early warning of the structure.

4.1 Principles of Computer Vision-Based Displacement Measurement

Computer vision has been widely applied in the field of civil engineering. Its basic principle is to simulate biological vision using computers and related electronic devices, whereby specific information required by users can be extracted from images or videos through dedicated algorithms. Based on practical engineering requirements, this study adopts a self-developed displacement measurement algorithm to accomplish target tracking and displacement monitoring. The theoretical implementation of this method consists of three main steps: camera calibration, target tracking, and three-dimensional reconstruction.

(1) Camera calibration

The purpose of camera calibration is to establish a one-to-one correspondence between image coordinates and real-world coordinates. For planar displacement measurement using a monocular camera, planar calibration methods can be employed. Common approaches include Zhang's calibration method and the scale factor method. In this chapter, the scale factor method is adopted, as it is a simpler and more efficient calibration approach. The principle of this method is to determine the ratio between a known actual length in the spatial coordinate system and the corresponding pixel length in the image coordinate system, and then convert the measured pixel displacement into actual displacement using this ratio. The scale factor is calculated as:

$$\text{scale} = \frac{d}{p} \tag{4.1}$$

where scale is the scale factor, d is the known actual distance between two points, and p is the corresponding pixel distance between the two points. When the optical axis of the camera is perpendicular to the plane of motion of the target, this method provides high calibration accuracy and is particularly feasible in scenarios where personnel access or manual operation is difficult.

(2) Target tracking

Target tracking is a key step in displacement measurement, as the accuracy and precision of target tracking directly determine the accuracy of displacement measurement. In this chapter, a template matching technique [101, 102] is adopted for target tracking. The basic principle is illustrated in Fig. 4.1. First, the position of a predefined marker is tracked using the template matching method, and then the center corner point of the marker is identified as the precise target point for displacement measurement.

The template matching method is an algorithm used to search for the region in an image that best matches a given template. Its tracking principle is illustrated in Fig. 1a. First, a tracking target is defined in a reference image using a rectangular window, which is referred to as the predefined template. Then, a rectangular search window of the same size is moved pixel by pixel within the region of interest (ROI) of subsequent images, and the similarity at each position is calculated in order to identify the best matching location. The degree of matching between the search window and the predefined template is described by the normalized correlation coefficient:

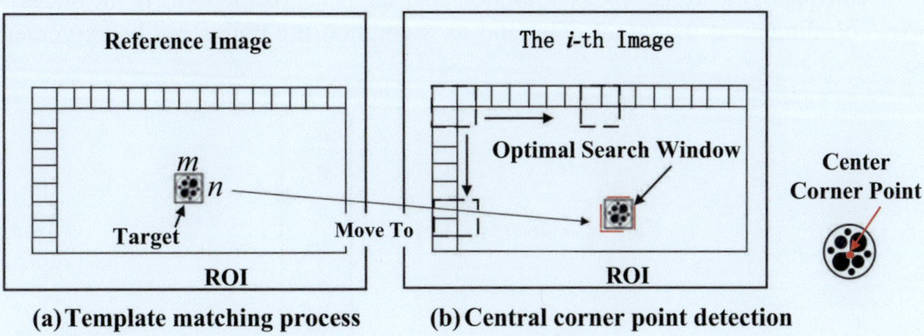

Fig. 4.1 Recognition principle of the template-corner algorithm

$$c(i,j) = \frac{\sum_{x=0}^{m-1}\sum_{y=0}^{n-1}(f(x,y)-\overline{f})(g_t(x+i,y+j)-\overline{g}_t)}{\sqrt{\left[\sum_{x=0}^{m-1}\sum_{y=0}^{n-1}(f(x,y)-\overline{f})\right]\left[\sum_{x=0}^{m-1}\sum_{y=0}^{n-1}(g_t(x+i,y+j)-\overline{g}_t)\right]}} \tag{4.2}$$

where c is the correlation coefficient at a given time instant($-1 < c < 1$);m and n are the length and width of the search window and the predefined template, respectively; $f(x, y)$ denotes the pixel intensity at a point in the predefined template, $\overline{f}$ is the mean gray value of the predefined template; $g_t(x,y)$ denotes the pixel intensity at a point in the search window at time t, and $\overline{g}_t$ is the mean gray value of the search window at time t. The closer the value of c is to 1, the more likely the corresponding position represents the target location after motion. A feature region at a certain moment before monitoring begins is selected as the predefined template, and for each subsequent image during monitoring, Eq. (4.2) is used for calculation. The region corresponding to the value of c closest to 1 in the image is identified as the target position at that moment.

Since the template matching method can only obtain integer-pixel coordinates, the measurement accuracy is limited. Therefore, subpixel techniques are required to further improve the measurement precision. In this study, a bi-parabolic fitting method is adopted to achieve subpixel displacement measurement accuracy [103]. It is assumed that the distribution of the normalized correlation coefficient c in the vicinity of the subpixel coordinates of the target point can be represented by a bi-parabolic function:

$$c(x, y) = a_0 + a_1x + a_2y + a_3x^2 + a_4xy + a_5y^2 \tag{4.3}$$

where a_0, a_1, a_2, a_3, a_4, and a_5 are coefficients to be determined.

Accordingly, the normalized correlation coefficients corresponding to the integer-pixel coordinates of the target point and its 8 neighboring points can be expressed as:

$$c_{ij} = \begin{bmatrix} 1 & x_i & y_j & x_i^2 & x_iy_j & y_j^2 \end{bmatrix} \begin{bmatrix} a_0 \\ a_1 \\ a_2 \\ a_3 \\ a_4 \\ a_5 \end{bmatrix} \tag{4.4}$$

where $i, j = -1, 0, 1$, and c_{00} denotes the normalized correlation coefficient at the integer-pixel coordinate of the target point. This formulation yields a system of equations containing nine independent equations. The 6 unknown coefficients can be solved using the least-squares method. Substituting these coefficients back into Eq. (4.4) yields the fitted function, and the coordinates corresponding to the maximum value of this function are taken as the subpixel coordinates (x_{sub}, y_{sub}):

$$\begin{cases} \frac{\partial c(x,y)}{\partial x} = a_1 + 2a_3x + a_4y = 0 \\ \frac{\partial c(x,y)}{\partial y} = a_2 + a_4x + 2a_5y = 0 \end{cases} \Rightarrow \begin{cases} x_{sub} = \frac{2a_1a_5 - a_2a_4}{a_4^2 - 4a_3a_5} \\ y_{sub} = \frac{2a_2a_3 - a_1a_4}{a_4^2 - 4a_3a_5} \end{cases} \tag{4.5}$$

In template matching, the target region detected in the i-th image is denoted as A_i. Furthermore, a corner detection algorithm can be applied within the detected region to identify the location of the target corner point:

$$\mathbf{C} = \sum_{x,y} \left\{ exp[-(x^2 + y^2)/40] \begin{bmatrix} \left(\frac{\partial I}{\partial x}\right)^2 & \frac{\partial I}{\partial x}\frac{\partial I}{\partial y} \\ \frac{\partial I}{\partial x}\frac{\partial I}{\partial y} & \left(\frac{\partial I}{\partial y}\right)^2 \end{bmatrix} \right\} \tag{4.6}$$

where $\mathbf{C}$ denotes a 5 × 5pixel subwindow within the detected region $\mathbf{A}_i$; (x, y) represents the pixel coordinates of a point within the 5 × 5 subwindow; I is the pixel intensity at (x,y); and λ_1 and λ_2 are the two eigenvalues of $\mathbf{C}(\lambda_1 < \lambda_2$. By searching all subwindows within the detected region $\mathbf{A}_i$ and identifying the subwindow with the minimum λ_1, the center of that subwindow is determined as the target corner point. In the proposed method, the center corner point of the marker identified is shown in Fig. 4.1b.

The subpixel coordinates of the monitoring point are obtained by solving the minimum of the following expressions to further improve coordinate accuracy:

$$\text{Min}\left\| \sum_{\hat{x},\hat{y}} w_{\hat{x},\hat{y}} \left(\left(\begin{bmatrix} x \\ y \end{bmatrix} - \begin{bmatrix} \hat{x} \\ \hat{y} \end{bmatrix} \right)^T (\nabla I_{\hat{x},\hat{y}})(\nabla I_{\hat{x},\hat{y}})^T \left(\begin{bmatrix} x \\ y \end{bmatrix} - \begin{bmatrix} \hat{x} \\ \hat{y} \end{bmatrix} \right) \right) \right\| \tag{4.7}$$

$$\nabla I_{\hat{x},\hat{y}} = \left[\frac{\partial I}{\partial \hat{x}} \ \frac{\partial I}{\partial \hat{y}} \right]^T \tag{4.8}$$

where (x, y) are the precise subpixel coordinates of the target corner point in the image; (x_0, y_0) are the integer-pixel coordinates of the target corner point obtained from Eq. (4.6); and ($\hat{x}$, $\hat{y}$) are the neighboring integer-pixel coordinates around (x_0, y_0). The subpixel coordinates of the target corner point can be obtained through least-squares fitting.

(3) Three-dimensional reconstruction

Once the positions of the monitoring points at different time instants in the image are determined, the image coordinates can be inversely transformed using the system calibration information (Eq. (4.9)) to obtain the corresponding positions of the monitoring points in the real-world coordinate system. Since the scale factor method is adopted for calibration, the transformation does not consider camera orientation or intrinsic parameters. Therefore, the parameter matrix $\mathbf{P}$ is an identity matrix, and Eq. (4.9) degenerates into Eq. (4.10).

$$\mathbf{M} = s\mathbf{P}^{-1}\mathbf{m} \tag{4.9}$$

$$\mathbf{M} = s\mathbf{m} \tag{4.10}$$

Let the real-world position coordinates of n monitoring points at the initial time be $\begin{bmatrix} x_{1-0}\ x_{2-0}\ \cdots\ x_{n-0} \\ y_{1-0}\ y_{2-0}\ \cdots\ y_{n-0} \end{bmatrix}^T$, and the position coordinates at time t be $\begin{bmatrix} x_{1-t}\ x_{2-t}\ \cdots\ x_{n-t} \\ y_{1-t}\ y_{2-t}\ \cdots\ y_{n-t} \end{bmatrix}$. Then, the displacement of these points can be expressed as:

$$\begin{bmatrix} \Delta x_{1-t} & \Delta y_{1-t} \\ \Delta x_{2-t} & \Delta y_{2-t} \\ \vdots & \\ \Delta x_{n-t} & \Delta y_{n-t} \end{bmatrix} = \begin{bmatrix} x_{1-t} & y_{1-t} \\ x_{2-t} & y_{2-t} \\ \vdots & \\ x_{n-t} & y_{n-t} \end{bmatrix} - \begin{bmatrix} x_{1-0} & y_{1-0} \\ x_{2-0} & y_{2-0} \\ \vdots & \\ x_{n-0} & y_{n-0} \end{bmatrix} \tag{4.11}$$

There are two main limitations in the traditional VVM method described above. First, in order to improve measurement accuracy under flexible camera viewpoints, the VVM method employs a template-corner approach for target tracking. However, the template matching method is not suitable for tracking targets with rotational components and may lead to significant matching errors. Second, the corner detection method used in this approach is sensitive to image quality. It generally performs well in indoor experiments where environmental conditions are stable, and the captured images have high quality, which is favorable for accurate computer vision measurements. In outdoor environments, insufficient illumination or fog can cause failures in corner detection. To address the issues of incomplete markers, blurred images, insufficient accuracy, and low robustness under adverse conditions, an improved algorithm (DAVIM) is proposed. When the target undergoes rigid-body rotation, a CNN-based detector (YOLOv4) is used to replace the template matching method for recognizing occluded markers (Fig. 4.2) [104]. When image quality is degraded by harsh environmental conditions, super-resolution generative adversarial networks (SRGAN) are employed to enhance image resolution (Fig. 4.3) [103, 105]. However, in practical monitoring, it is observed that external environmental vibrations may cause camera defocusing, leading to difficulties in target capture. Therefore, this chapter proposes a motion-adaptive visual measurement method (M-DAVIM) that integrates dynamic target tracking with time–frequency joint correction, and its workflow is shown in Fig. 4.4. Based on DAVIM, this method applies an affine transformation model to correct vibration signals in the time domain, and then uses variational mode decomposition (VMD) for frequency-domain correction to separate object vibration and camera vibration from the original signal [106]. The affine transformation model is a fundamental linear geometric mapping that preserves collinearity and parallelism. Owing to its concise matrix representation, which can uniformly

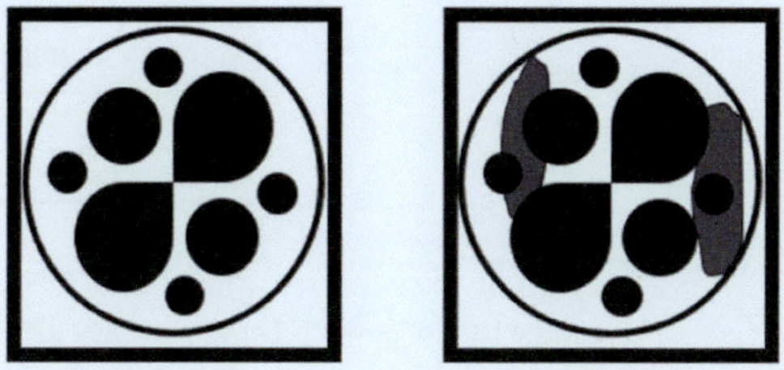

Fig. 4.2 YOLOX-based recognition of occluded markers

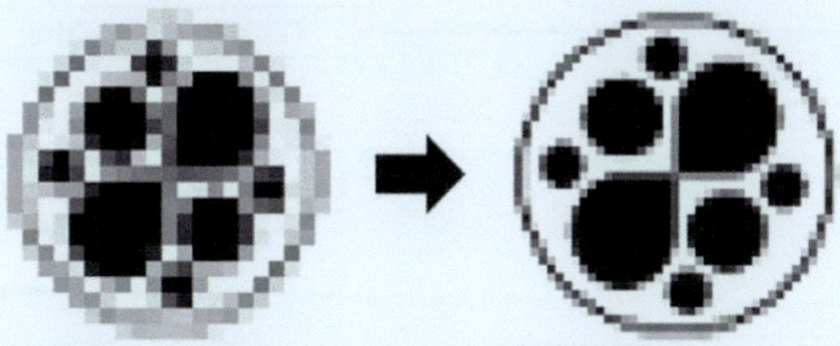

Fig. 4.3 Resolution enhancement using the SRGAN network

describe translation, rotation, scaling, and shearing, it is widely used in image registration and motion estimation. Variational mode decomposition (VMD) is a technique that decomposes a given signal into a finite number of sub-signals or modes while adaptively determining their center frequencies. Through a multimodal data processing framework, the proposed M-DAVIM method effectively mitigates the accuracy degradation caused by camera motion, environmental noise, and target deformation in traditional visual measurement approaches.

4.2 Influence of External Factors on Measurement Accuracy

4.2.1 Major External Influencing Factors

Computer vision-based measurement technology has been widely applied in the field of civil engineering due to its non-contact nature, high efficiency, and high degree of automation. However, its measurement accuracy is highly susceptible to interference from various external factors. Five common key external variables that affect the accuracy of computer vision-based measurements include illumination intensity, fog concentration, camera elevation angle, marker size, and camera vibration induced by the external environment.

(1) Illumination intensity

In visual systems, variations in illumination constitute a fundamental challenge to image quality. Insufficient lighting, overexposure, or non-uniform illumination can directly interfere with the accurate recording of scene information, leading to issues

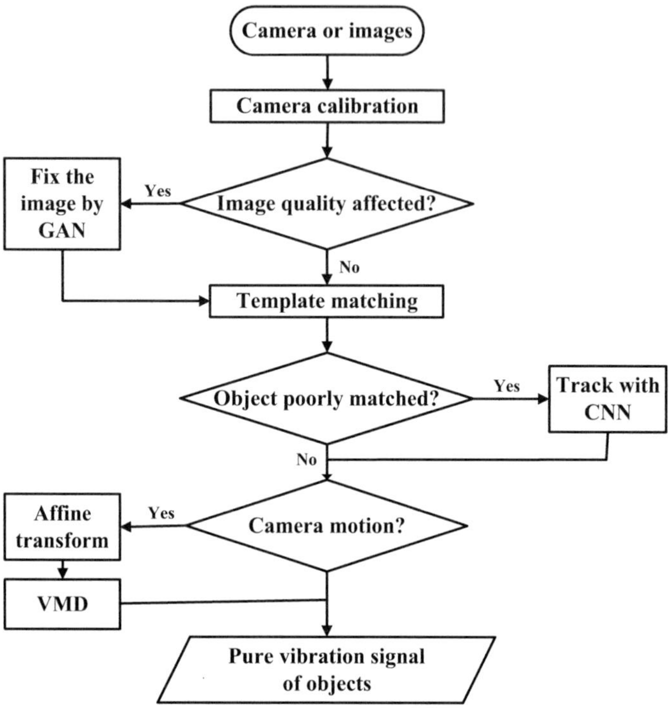

Fig. 4.4 Workflow of the M-DAVIM method

such as feature point loss, reduced contrast, or local image distortion. These effects weaken or distort critical feature information in the image, thereby degrading the performance of target tracking algorithms or even causing their failure.

In laboratory environments, illumination is typically provided by controllable artificial light sources, whose intensity and direction can be precisely adjusted, and whose distribution is uniform and stable. Such conditions are ideal for maintaining high measurement accuracy. In contrast, outdoor engineering applications mainly rely on natural sunlight, whose intensity, angle, and spectral characteristics vary dynamically with time and weather conditions. The instability of outdoor illumination often results in complex scenarios such as specular highlights, shadowed regions, or backlighting in images, posing severe challenges to the robustness of computer vision algorithms and, in some cases, directly leading to algorithm failure.

(2) Fog concentration

In a broad sense, fog refers to aerosol systems composed of suspended water droplets, ice crystals, dust, or smoke particles in the atmosphere. In visual measurement, fog primarily interferes with optical signal transmission through atmospheric scattering effects, which manifest as increased image noise, shifts in grayscale distribution, and a significant reduction in contrast. These degradation effects severely disrupt visual tasks such as feature extraction, edge detection, and target matching, thereby affecting subsequent localization, recognition, and measurement accuracy. In particular, under severe weather conditions, dense fog can obscure target textures and contours, causing tracking algorithms to lose the target or exhibit drift, making reliable target tracking difficult.

(3) Camera pose

Although Zhang's camera calibration method and other projection geometry-based calibration approaches can accommodate flexible camera placement, the pose of the camera relative to the target still affects the appearance of target features in the image and the relative range of target motion within the image. Consequently, camera pose can influence the performance of vision algorithms. During outdoor monitoring, variations in the camera elevation angle may introduce perspective distortion of the target, which adversely affects the accuracy of feature point extraction and displacement estimation.

(4) Target size

In general, larger targets in an image exhibit more prominent features, which are more conducive to recognition by computer vision algorithms. Moreover, for a given physical size, a larger target occupies more pixels in the image, resulting in a smaller real-world dimension per pixel and, consequently, a higher achievable theoretical measurement accuracy. However, in practical applications, the target size cannot be increased indefinitely due to constraints such as site space, lens focal length limitations, and the requirements of multi-point synchronous measurement. Therefore, the selection of target size requires a balance between experimental conditions and measurement accuracy.

(5) Camera vibration

In outdoor construction environments, frequent movement of concrete mixer trucks and engineering vehicles, as well as continuous operation of heavy machinery (e.g., pile drivers, mixers, and cranes), inevitably induces multi-frequency composite vibrations. These vibrations are transmitted to the monitoring camera through multiple paths, causing additional camera motion. The coupling between camera motion and structural dynamic response results in displacement signals extracted from image sequences being composite signals that include camera motion components, rather than pure structural responses, thereby adversely affecting the accuracy of structural response measurements.

4.2.2 *Experimental Design and Analysis of Factor Impacts on Accuracy*

To systematically investigate the influence of different external factors on the accuracy of the computer vision-based measurement method proposed in this chapter, this section analyzes and summarizes the variation patterns of measurement errors with reference to the main external factors listed in Sect. 4.2.1, namely illumination intensity, fog concentration, camera pose, target size, and camera vibration. To obtain evaluation results with practical comparability, all measurement errors in the experiments are uniformly converted into measurement errors expressed in millimeters (mm) using the scale factor. The instrument models and their main technical parameters used in the experiments are listed in Table 4.1.

(1) Illumination Intensity

The experimental setup for the illumination tests is shown in Fig. 4.5. Using a controllable light source and an illuminance meter, the illumination intensity can be adjusted within a range of 10–600 lx. A dual-sensor comparison scheme is adopted for displacement measurement: a high-precision laser displacement sensor (LDS) is used as the reference benchmark, while a camera simultaneously captures images of a predefined marker attached to the motor. The experimental procedure consists

Table 4.1 Testing instruments and their main technical parameters

Instrument	Model	Main parameters
Swing motor	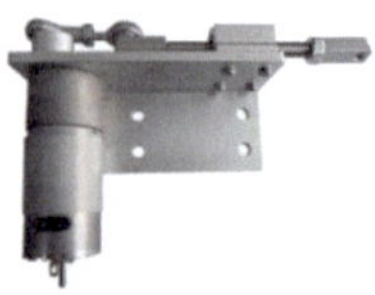	Voltage: DC 12/24 V No-load current: 0.02–0.13 A Rotational speed: 5–500 rpm Stroke: 28 mm
Laser displacement sensor		Measurement range: 40–70 mm Repeatability: 5 μm Linearity: ± 0.1% of full scale Sampling frequency: 100 Hz
Camera		Xiaomi 13 smartphone Image sensor: Sony IMX800 Maximum resolution: 8192 × 6144 Pixel size: 1.0 μm Color mode: RGB
Illuminance meter		Measurement range: 0–200,000 lx Sampling frequency: 2 Hz Accuracy: ± 3% ± 10 digits
Controllable light source		Input voltage/frequency: 240 V/50Hz Rated current: 0.07 A Power: 60 W

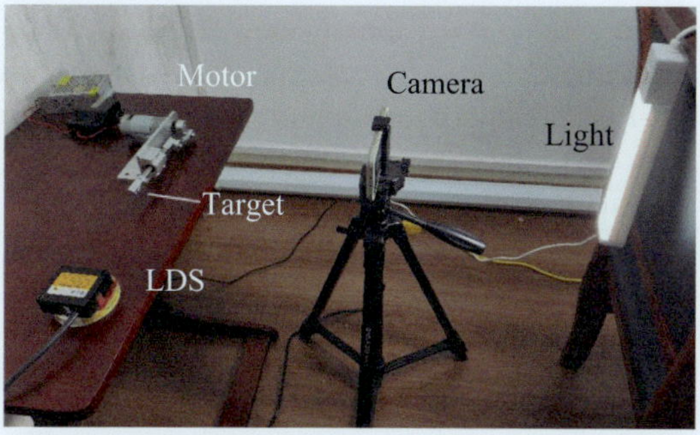

Fig. 4.5 On-site illumination experiment setup

of three consecutive steps: (i) calibrating and activating the motor drive system to generate displacement; (ii) synchronously acquiring displacement data from the LDS and video streams from the camera; and (iii) extracting the marker displacement from the video sequence using the M-DAVIM algorithm. Under different illumination conditions, the deviation between the vision-based displacement measurements and the LDS reference values is quantified, thereby enabling a quantitative evaluation of the influence of illumination variation on the accuracy of vision-based displacement measurement.

A total of five graded illumination levels were set in this experiment: 10 lx, 50 lx, 200 lx, 400 lx, and 600 lx, to investigate the influence of lighting conditions on the performance of the M-DAVIM system. As shown in Fig. 4.6, images acquired under these different illumination levels are presented sequentially (from left to right: 10 lx, 50 lx, 200 lx, 400 lx, and 600 lx). It can be observed that as the illumination intensity decreases from 600 to 10 lx, the image quality deteriorates significantly. Specifically, the global average grayscale value decreases monotonically, while the noise level increases markedly.

To quantitatively evaluate the influence of illumination intensity on the accuracy of the vision-based measurement system, this study uses high-precision sensor data as the reference benchmark to analyze the vision-based displacement measurements obtained under different lighting conditions, and the root mean square error (RMSE) and mean absolute error (MAE) are calculated. The results, as shown in Fig. 4.7, indicate that under low illumination conditions (10 lx and 50 lx), both RMSE and MAE increase significantly; in particular, the RMSE reaches 1.1 mm, which is 2.1

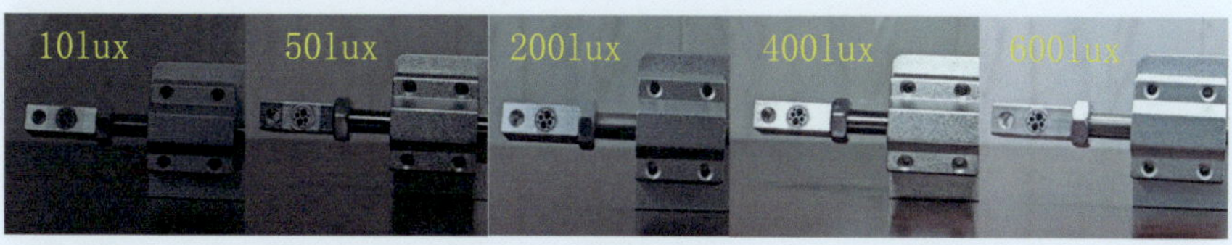

Fig. 4.6 Images under different illumination intensities

times the corresponding value at an illumination level of 400 lx (0.51 mm), and the maximum error reaches 5 mm, demonstrating that measurement errors increase markedly under low-light conditions. Within the illumination range of 200–400 lx, RMSE and MAE decrease to their lowest levels, indicating that this range represents the optimal operating interval for measurement. In contrast, when the illumination intensity is further increased to 600 lx, RMSE and MAE exhibit an increasing trend again, accompanied by a reduction in measurement accuracy, with the maximum error increasing to approximately 4 mm. Therefore, to ensure high accuracy and reliability of vision-based displacement measurements, the ambient illumination intensity should be controlled within the optimal range of 200–400 lx, and monitoring under insufficient or excessively strong lighting conditions should be avoided.

(2) Fog Concentration

In this study, experimental scenarios with five graded fog concentration levels (light, light-moderate, moderate, moderate-heavy, and heavy) were established. During the experiments, a humidifier was used to control the fog concentration, while vision-based displacement data and LDS displacement data were synchronously collected to systematically evaluate the influence of fog on the M-DAVIM vision system. The on-site fog experiment setup is shown in Fig. 4.8.

Figure 4.9 presents the images of the feature target under the five fog conditions. Light fog causes slight background blurring; moderate fog induces a halo effect around the target contours; and heavy fog reduces the contrast between the target and the background to nearly zero. From the images, it can be observed that increasing fog concentration monotonically raises the average background grayscale value and significantly aggravates image blurring.

Figure 4.10 illustrates the variation trends of RMSE and MAE with fog concentration. The results indicate a negative correlation between fog concentration and vision-based measurement accuracy, with measurement errors increasing exponentially as fog concentration increases. As fog conditions transition from light to moderate and then to heavy fog, both RMSE and MAE continuously increase, accompanied

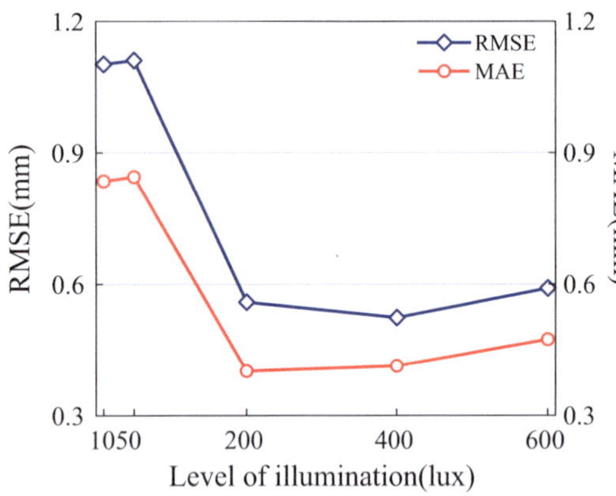

Fig. 4.7 Comparison of measurement errors under different illumination intensities

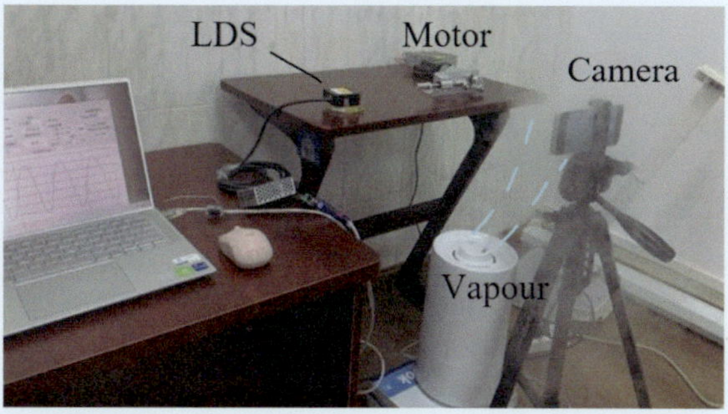

Fig. 4.8 Fog experiment setup

Fig. 4.9 Images under different fog concentrations

by a systematic degradation in measurement accuracy. These findings confirm the significant limitations of the M-DAVIM vision system under heavy fog conditions; therefore, severe fog environments should be avoided in practical deployments.

(3) Camera Elevation Angle

To systematically investigate the influence of camera elevation angle on measurement accuracy, experimental conditions with different elevation angles (5°, 10°, 15°, 20°, 25°, and 30°) were designed. The elevation angle was adjusted by modifying the

Fig. 4.10 Comparison of measurement errors under different fog concentrations

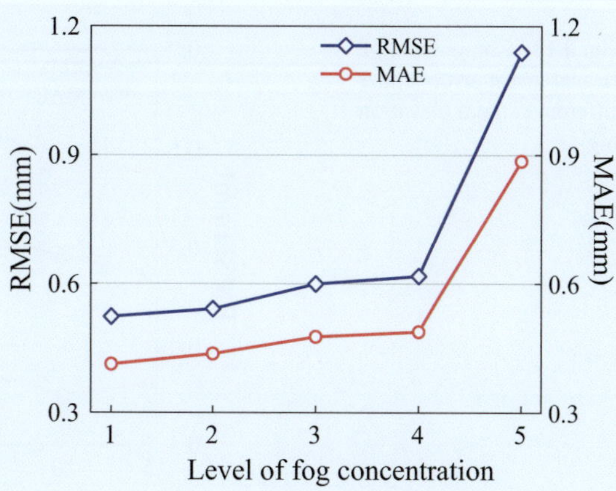

Fig. 4.11 Camera angle test setup

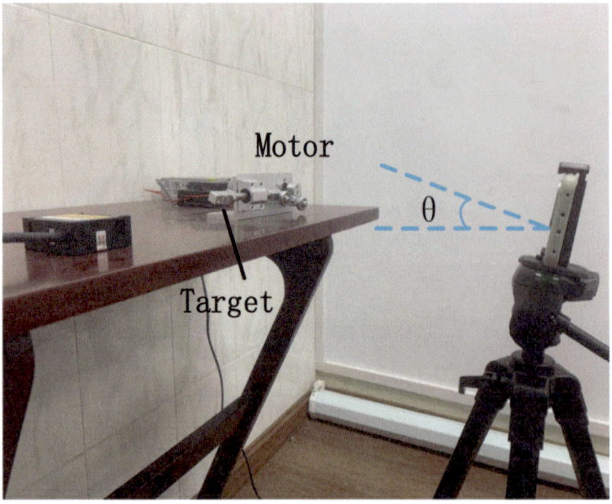

tripod configuration, and the effects of different camera elevation angles on the M-DAVIM vision system were evaluated by comparing vision-based measurements with LDS data. The experimental setup is shown in Fig. 4.11.

Figure 4.12 presents the comparison of measurement errors under different camera elevation angles. As the camera elevation angle increases from 5° to 30°, both RMSE and MAE exhibit a gradual increasing trend, although the growth rate slows significantly within the range of 25°–30°. At an elevation angle of 5°, the RMSE and MAE are 0.47 mm and 0.36 mm, respectively. When the elevation angle increases to 30°, the RMSE rises to 0.64 mm (an increase of 36.2% compared with 5°), and the MAE reaches 0.48 mm (an increase of 33.3%). Both error metrics remain below 1 mm, indicating that although larger elevation angles increase measurement errors, the M-DAVIM system consistently maintains sub-millimeter accuracy within the 5°–30° range, satisfying the requirements for outdoor monitoring.

Fig. 4.12 Comparison of measurement errors under different camera elevation angles

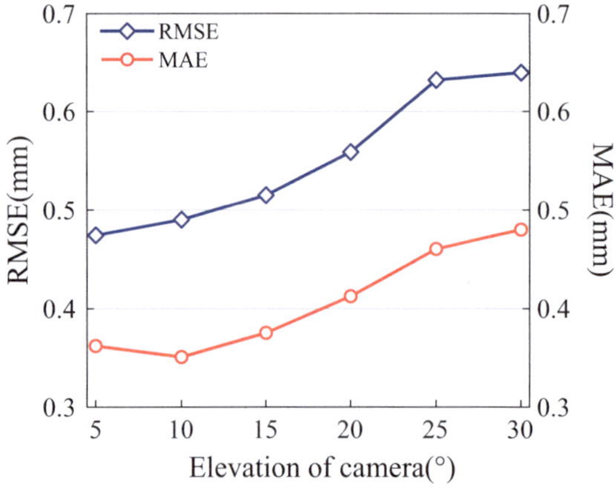

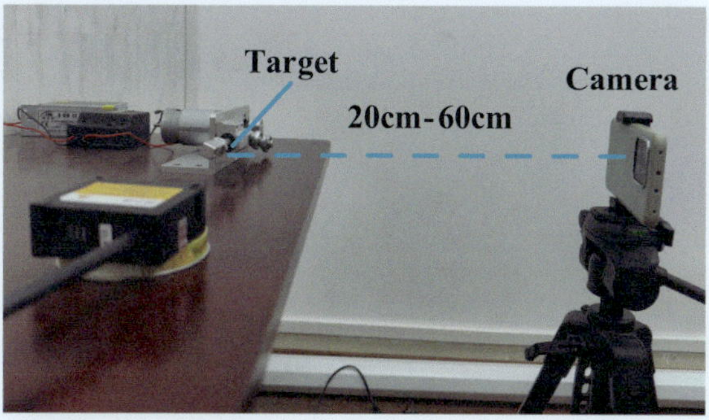

Fig. 4.13 Experimental setup for different camera-target distances

(4) Target Size

In this study, the pixel size of the marker on the image plane was controlled by systematically adjusting the distance between the camera and the marker. The experimental setup is shown in Fig. 4.13. Using LDS displacement measurements as the reference values, the vision-based displacement results obtained under different pixel sizes were compared to quantitatively evaluate the influence of marker imaging size on measurement accuracy.

A total of five experimental conditions were tested to investigate the effect of target size on measurement results. The camera-to-target distance varied from 20 to 60 cm, corresponding to marker pixel sizes of 30 pixels, 25 pixels, 18 pixels, 14 pixels, and 10 pixels, respectively, as shown in Fig. 4.14.

Figure 4.15 shows that both RMSE and MAE of displacement measurements increase monotonically as the marker pixel size decreases. When the pixel size is greater than or equal to 30 pixels, the system achieves sub-pixel accuracy (RMSE $\leq$ 0.5 mm). When the pixel size falls below 18 pixels, both RMSE and MAE increase significantly, and the maximum error rises sharply from 2 mm at 18 pixels to 20 mm at 10 pixels. Therefore, for outdoor deployment of the M-DAVIM system, the marker imaging size should be maintained above 18 pixels to prevent systematic accuracy degradation caused by excessively small targets.

(5) Camera Vibration

During outdoor operations, construction vehicles and heavy machinery operate continuously, inevitably inducing multi-frequency composite vibrations. The vibration spectrum at construction sites typically ranges from 5 to 50 Hz. To investigate

Fig. 4.14 Marker pixel sizes under different distances

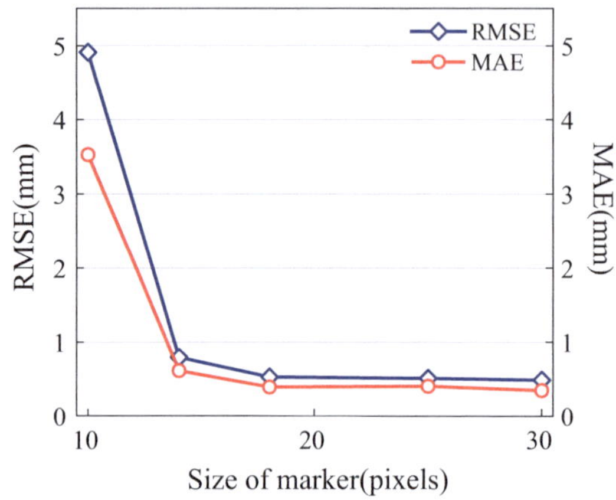

Fig. 4.15 Comparison of measurement errors for different pixel sizes

the influence of site-induced vibrations on the accuracy of vision-based monitoring systems, a frequency-adjustable three-axis vibration table was used to simulate camera vibrations, as shown in Fig. 4.16. Both horizontal and vertical vibration conditions were examined. Using LDS data as the reference, the deviations between vision-based displacement measurements and reference values under different vibration frequencies were quantified. A comparative analysis between the DAVIM and M-DAVIM algorithms was conducted to verify the superiority of M-DAVIM under complex vibration environments. Figure 4.17 shows images captured under vibration frequencies of 5 Hz and 50 Hz, where significant image blurring can be observed at 50 Hz.

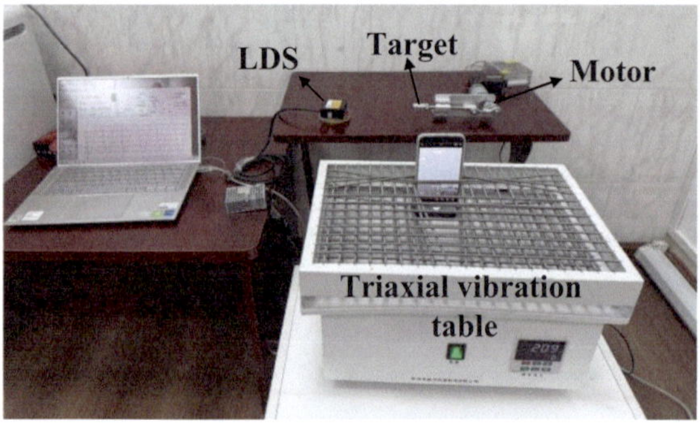

Fig. 4.16 Camera vibration experiment setup

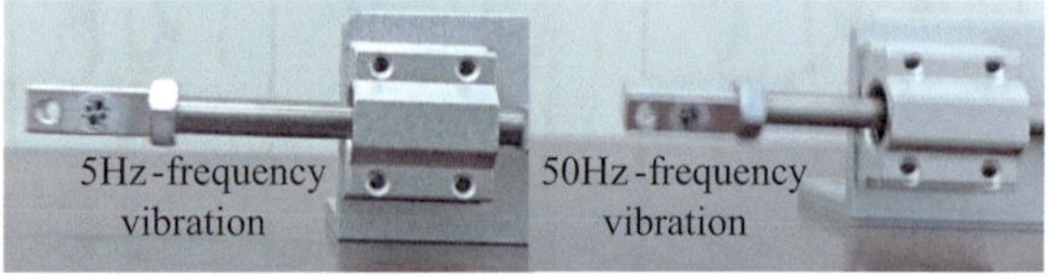

Fig. 4.17 Images under camera vibration

To investigate the effect of horizontal vibration on measurement accuracy, representative low-, medium-, and high-frequency conditions (5, 20, 35, and 50 Hz) were selected. Figure 4.18 compares displacement measurements obtained from LDS, DAVIM, and M-DAVIM. The results indicate that under low-frequency vibration (5 Hz), both DAVIM and M-DAVIM can accurately track displacement. However, as vibration frequency increases, the traditional DAVIM method exhibits systematic failure in the mid-to-high frequency range, showing significant phase shifts at 35 Hz and severe dynamic misalignment at 50 Hz. In contrast, M-DAVIM effectively eliminates interference through adaptive motion decoupling and frequency-domain filtering, maintaining high trajectory consistency at 35 Hz and relatively small errors at 50 Hz. These results demonstrate that the M-DAVIM system can achieve high-precision monitoring within the 50 Hz vibration range.

Figure 4.19 presents the comparison of displacement errors between DAVIM and M-DAVIM over the horizontal vibration frequency range of 5–50 Hz, quantified using RMSE and MAE. As vibration frequency increases, the displacement errors of the DAVIM method exhibit a two-stage growth pattern: in the low-to-medium frequency range (5–35 Hz), errors increase gradually (RMSE from 0.5 mm to 1.2 mm and MAE from 0.44 mm to 0.88 mm); in the high-frequency range (35–50 Hz),

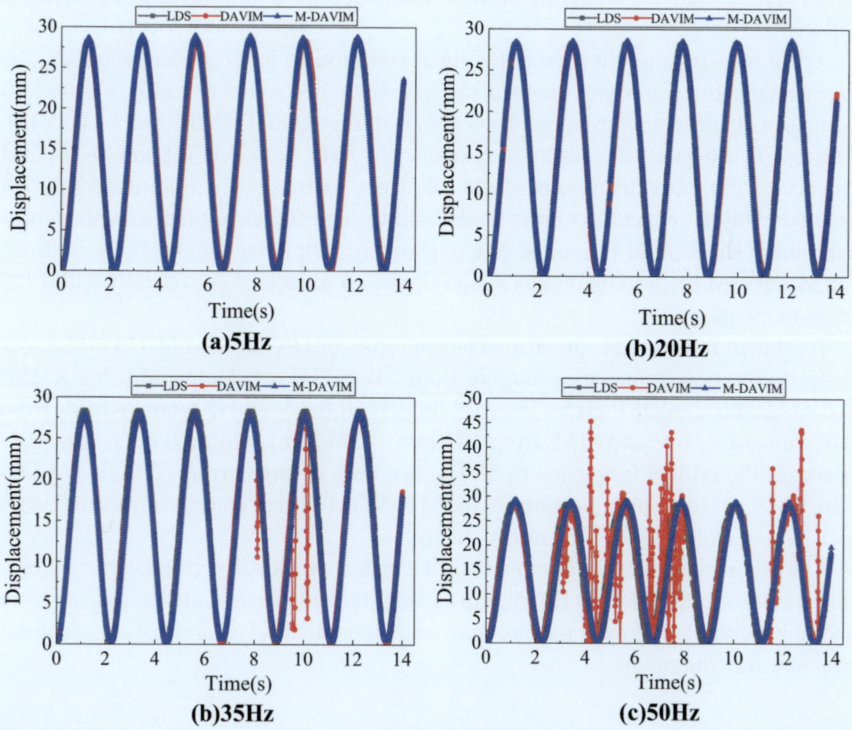

Fig. 4.18 Comparison of horizontal displacement measurements by LDS, DAVIM, and M-DAVIM

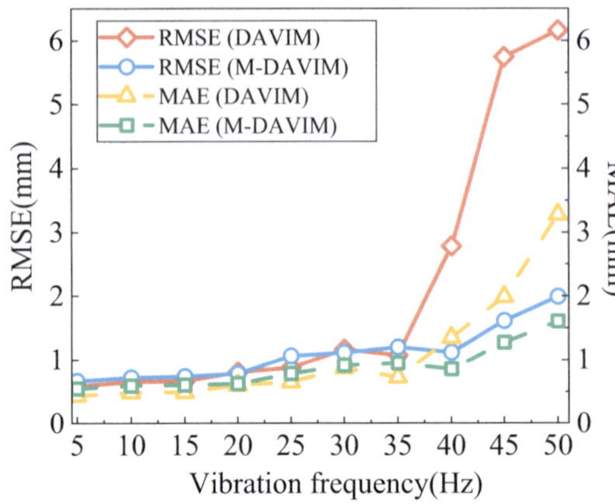

Fig. 4.19 Comparison of horizontal measurement errors under different vibration frequencies

the system becomes unstable, reaching peak errors at 50 Hz (RMSE = 6.1 mm, MAE = 3.2 mm). In contrast, the M-DAVIM system maintains strong robustness across the entire frequency range, with RMSE < 1.2 mm and MAE < 0.92 mm in the 5–35 Hz range, and significantly reduced errors at 50 Hz (RMSE = 1.433 mm, MAE = 1.125 mm), effectively resolving the high-frequency vibration-induced measurement failure.

When investigating the effect of vertical vibration on measurement accuracy, the experimental procedure was identical to that used for horizontal vibration. Figure 4.20 compares displacement measurements obtained from LDS, DAVIM, and M-DAVIM. The results show that at a vibration frequency of 5 Hz, both DAVIM and M-DAVIM can accurately track displacement. As frequency increases, the traditional DAVIM method exhibits reduced accuracy in the mid-to-high frequency range, with significant phase shifts at 30 Hz and severe dynamic misalignment at 50 Hz. In contrast, the M-DAVIM method maintains a high degree of trajectory consistency with LDS measurements.

As shown in Fig. 4.21, the displacement errors of DAVIM and M-DAVIM under vertical vibration frequencies ranging from 5 to 50 Hz are compared using RMSE and MAE. For DAVIM, errors increase linearly in the 5–45 Hz range (RMSE from 0.62 mm to 1.23 mm and MAE from 0.46 mm to 0.88 mm), while a sudden instability occurs at the critical frequency of 50 Hz, resulting in large errors (RMSE = 18.66 mm, MAE = 10.16 mm). In contrast, the M-DAVIM system controls errors to RMSE = 1.99 mm and MAE = 1.60 mm at 50 Hz.

The experimental results demonstrate that under both horizontal and vertical vibration conditions, the M-DAVIM system can significantly reduce measurement errors and substantially improve the accuracy and reliability of vision-based structural response measurements.

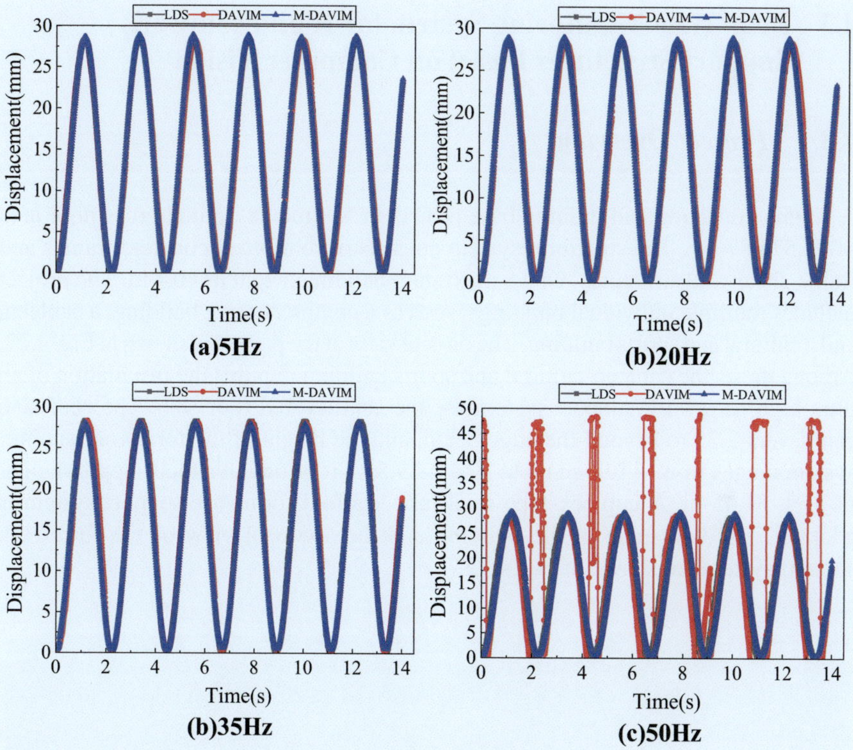

Fig. 4.20 Comparison of vertical displacement measurements by LDS, DAVIM, and M-DAVIM

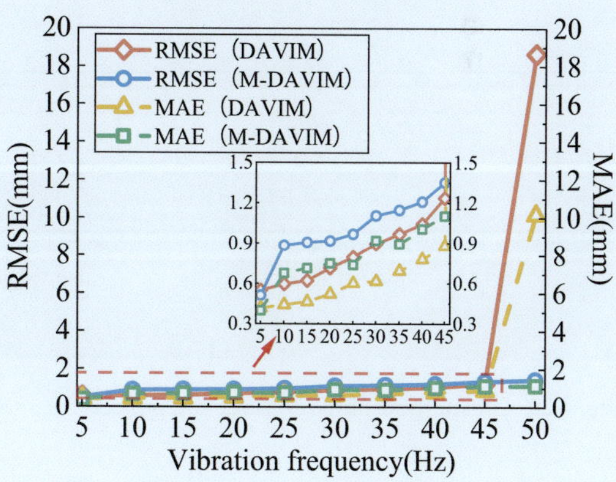

Fig. 4.21 Comparison of vertical measurement errors under different vibration frequencies

4.3 Real-Time Monitoring System for High Formwork Support Structures Based on Computer Vision

4.3.1 Project Overview

A certain education and training base project in Ningbo has a total gross floor area of 72,574.87 m^2. The structural system consists of reinforced concrete frames and frame-shear wall structures, with a maximum building height of 40.4 m. The project includes multiple individual buildings, such as a comprehensive building, a canteen, and a cultural and sports building. The on-site view of the project is shown in Fig. 4.22. Among these, the canteen-cultural and sports building requires the installation of an ultra-high formwork support system for the cantilevered roof slab. The roof slab cantilevers 2.25 m beyond the edge beam, and the height of the formwork support system ranges from −1.95 m up to 29.15–29.5 m, resulting in a total support height of 31.1–31.45 m. A cuplock-type steel tube scaffold formwork support system is adopted. The floor plan of the building and the sectional view of the formwork support system are shown in Fig. 4.23.

Fig. 4.22 On-site view of the construction project

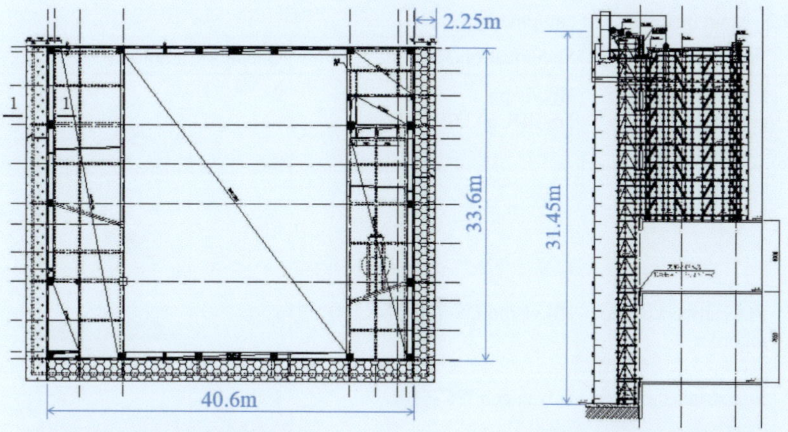

(a) Floor plan (b)Section 1-1 of the formwork support system

Fig. 4.23 Floor plan and sectional view of the canteen-cultural and sports building

4.3.2 Real-Time Displacement Monitoring During Concrete Pouring

An automated monitoring method based on computer vision is adopted to achieve real-time displacement monitoring during the concrete pouring process. Table 4.2 lists the main equipment and construction materials involved in this project.

Specific patterned markers are attached to the surface of the vertical posts using transparent adhesive tape within the field of view of the surveillance PTZ cameras. For the measurement points within the camera view, calibration parameters are obtained using the scale factor method. The displacement of the structural members can then be obtained directly through the real-time displacement monitoring software. As shown in Fig. 4.24, ① denotes the high formwork support system; ② denotes the markers attached to the vertical posts; ③ indicates the camera field of view; ④ represents the surveillance camera; and ⑤ shows the data analysis center and result display interface.

When deploying the monitoring system on site, the field of view and monitoring distance must satisfy the required measurement accuracy. The relationships among monitoring distance, focal length, field of view, and sensor size in the camera projection transformation can be expressed by the following equations:

$$\frac{D}{FOV} = \frac{f}{S} \tag{4.12}$$

$$FOV = Res \times \frac{n_r}{n_p} \tag{4.13}$$

Table 4.2 Main materials and equipment

S. No	Item name	Specification/Model	Quantity	Photo
1	PTZ surveillance camera	Hikvision DS-2DC6423IW-AE	2	
2	POEnetwork video recorder	Hikvision DS-7804N-K1/4P	1	
3	Industrial computer	Advantech IPC-610L	1	
4	Network switch	Hikvision DS-3E0524-E	1	
5	Network cable	Category 6 industrial Ethernet cable	100m	
6	Marker	/	As required	
7	Transparent adhesive tape	/	1	

By combining Eqs. (4.12) and (4.13), the following relationship is obtained:

$$n_r = n_p \cdot \frac{D}{f} \cdot \frac{S}{Res} \tag{4.14}$$

where D is the actual monitoring distance; FOV denotes the camera field of view; f is the camera focal length; S represents the camera sensor size; Res is the image resolution; n_r is the required measurement accuracy in the real world (mm); and n_p is the achievable measurement accuracy of the recognition algorithm (pixels).

In this study, the sensor size and imaging resolution of the surveillance cameras remain constant. The camera sensor size is 5.0 mm × 2.8 mm, and the image resolution is 1920 × 1088. Among the remaining parameters, the matching accuracy of the algorithm n_p is typically at the integer-pixel level (i.e., 1 pixel). However, when

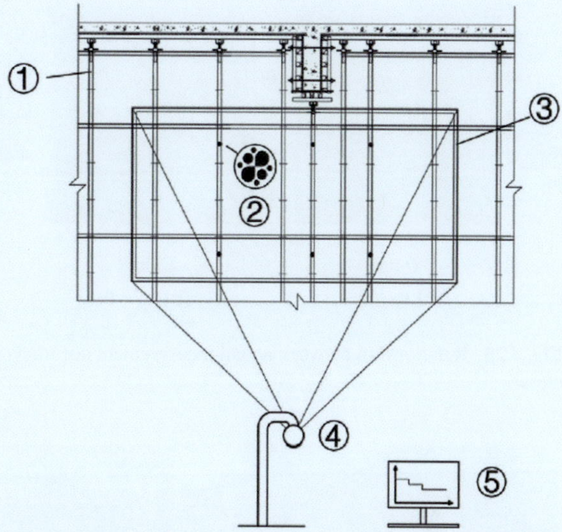

Fig. 4.24 Schematic diagram of the on-site monitoring system

subpixel matching techniques such as bi-parabolic fitting are adopted, the matching accuracy of target tracking can be improved to 0.02–0.05 pixels [105]. The focal length of the PTZ cameras used in this study can vary within the range of 4.8–110 mm. Substituting these parameters into the above equations yields an appropriate working distance for the cameras of 3.7–84.0 m. Figure 4.25 illustrates, in the form of contour plots, the relationships between the three variables and the achievable measurement accuracy in real scenarios. For example, in Fig. 4.25a, when the algorithm matching resolution is 1 pixel, achieving a measurement accuracy of 2 mm requires setting the monitoring distance and focal length according to the contour labeled "2". To obtain higher measurement accuracy in practical applications, one approach is to reduce the monitoring distance or increase the focal length, while another is to adopt subpixel matching techniques. As shown in Fig. 4.25b and c, when D and f remain constant, improving the matching resolution of the algorithm can effectively enhance measurement accuracy. These results provide guidance for configuring the monitoring system.

The monitored façade of the canteen-cultural and sports building is directly opposite the project office, with a straight-line distance of 40 m, which falls within the suitable operating range of the PTZ cameras. Therefore, the cameras are installed at the project office for monitoring. As shown in Fig. 4.26, the corner post and its adjacent left and right vertical posts are selected as monitoring targets. Two PTZ cameras are used to monitor regions at vertical intervals of 8 m along the height direction.

During on-site testing, due to installation constraints, it is sometimes necessary to slightly tilt the camera optical axis toward the object surface. According to the experimental results presented in Sect. 4.2.2, the measurement error of the vision-based monitoring system remains within an acceptable range for small tilt angles.

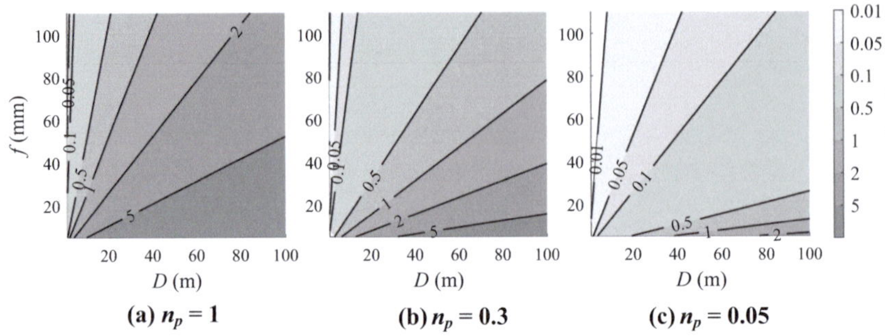

Fig. 4.25 Relationship between actual measurement accuracy(n_r) and monitoring system parameters

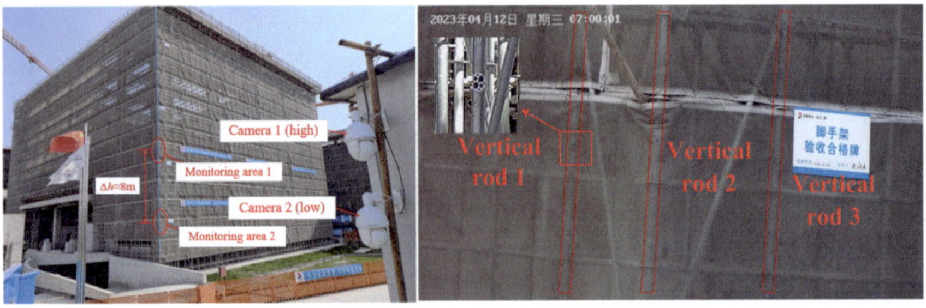

Fig. 4.26 Layout of the on-site monitoring system

The distance between Camera 2 and Monitoring Area 2 is 40 m with an elevation angle of 0°, while the distance between Camera 1 and Monitoring Area 1 is 40.8 m with an elevation angle of 11.3°. To obtain an appropriate monitoring range, the focal lengths of Camera 1 and Camera 2 are set to 49 mm and 64 mm, respectively.

The actual spacing between two adjacent disc couplers on a vertical post is 500 mm. According to Eq. (4.1), the scale factors for each measurement point are calibrated, and the results are listed in Table 4.3. This table establishes the transformation relationship between the image coordinate system and the real-world coordinate system. It should be noted that due to variations in the thickness of disc couplers, the top or bottom edge with a clear boundary should always be selected during calibration.

Since the surface of the marker is covered with a layer of transparent adhesive tape, partial occlusion may occur in some target regions. Therefore, when extracting the region of interest (ROI) and defining the predefined template, the image of the object covered by transparent tape should be used as the reference to ensure consistency between the template features and the tracked target. The monitoring software acquires the PTZ camera images in real time, and the interface displays the positions, IDs, and current displacement data of multiple measurement points simultaneously. When deformation exceeds a predefined threshold, the software triggers an alarm,

Table 4.3 Calibration coefficients of vertical posts

Vertical post ID	Actual distance (mm)	Pixel distance (pixels)	Calibration coefficient (mm/pixel)
Post 1 (upper)	500	228	2.193
Post 2 (upper)	500	236	2.119
Post 3 (upper)	500	232	2.155
Post 1 (lower)	500	306	1.634
Post 2 (lower)	500	312	1.603
Post 3 (lower)	500	308	1.623

prompting an inspection of abnormal conditions. If the rectangular tracking box deviates significantly from the target region, it is identified as a false match and false alarm, and the abnormal data should be excluded. Conversely, if the rectangular box accurately tracks the target, the construction process should be reviewed to investigate potential structural issues. In addition, when displacement data from multiple points on a single member are available, some data points below the alarm threshold may still contain noise. In such cases, the average displacement of these points is used to determine the member displacement, thereby minimizing the influence of noise. Ultimately, displacement–time histories of all members throughout the construction process can be obtained. After importing the calibration file into the monitoring program, real-time monitoring begins. The real-time displacement monitoring interface is shown in Fig. 4.27, where the positions, IDs, and current displacements of multiple measurement points are displayed. For example, in the view of Camera 1 monitoring the upper region, "No.1 (3.24, 1.44)" indicates that the horizontal displacement of measurement point 1 is 3.24 mm and the vertical displacement is 1.44 mm, where horizontal displacement is defined as positive to the right and vertical displacement as positive upward.

(a) Camera 1 view **(b) Camera 2 view**

Fig. 4.27 On-site monitoring system interface

4.3.3 Monitoring Data Analysis

Taking the daytime monitoring data of a representative day (27 May 2023) as an example, the displacement time-history curves and inclination time-history curves of the vertical posts from 7:00 to 17:00 are plotted, as shown in Fig. 4.28. The initial reference state for this monitoring is selected as the image recorded at 7:00 on 22 April. The following conclusions can be drawn from the results:

(1) The displacement and inclination of the vertical posts vary only slightly throughout the day and generally fluctuate around their mean values. Different vertical posts exhibit different mean displacements and inclinations in different regions and directions, which can be attributed to cumulative deformation effects resulting from long-term construction activities.
(2) During the period from 7:00 to 17:00 on that day, the horizontal displacement of the steel pipe support vertical posts ranges from −2 mm to 4 mm, the vertical displacement ranges from −3 mm to 4 mm, and the inclination ranges from 0 to 0.5‰. All these values are below the alarm thresholds specified in the Technical Specification for Automated Monitoring of Steel Pipe Support Scaffolding in Construction (DBJ33/T1296-2023), namely 12 mm for horizontal displacement, 8 mm for settlement, and 4‰ for inclination. None of the indicators exceeded the alarm limits, indicating that the steel pipe support system remained in a safe operating condition throughout the day.
(3) The maximum inclination of the vertical posts on that day is approximately one-eighth of the alarm threshold, indicating that inclination is not the controlling factor in vertical post monitoring. In contrast, horizontal and vertical displacements are of the same order of magnitude as their respective alarm limits and therefore require primary attention during monitoring.

4.4 Summary

This chapter mainly investigates the influence of different external factors on the proposed M-DAVIM method and its practical application in monitoring systems. After summarizing common external influencing factors, a series of controlled vibration experiments were conducted to systematically study the effects of illumination intensity, fog concentration, camera pose, target size, and camera vibration on measurement accuracy. A laser displacement sensor was used as the reference to quantitatively evaluate measurement errors under different external conditions. In addition, a vision-based monitoring system for a high formwork support structure was deployed at a construction site in Ningbo, enabling real-time tracking and displacement monitoring of key nodes of the target members. Based on the research results, the main conclusions are summarized as follows:

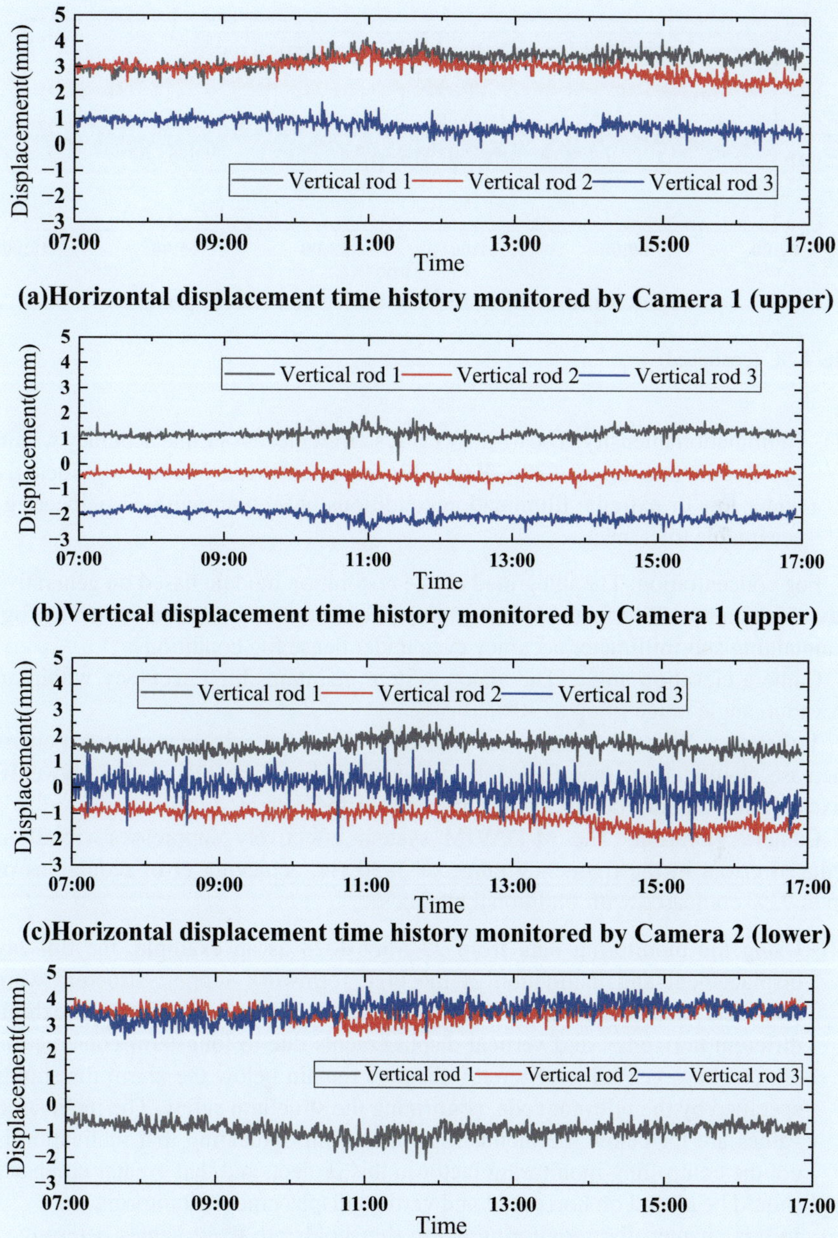

(a)Horizontal displacement time history monitored by Camera 1 (upper)

(b)Vertical displacement time history monitored by Camera 1 (upper)

(c)Horizontal displacement time history monitored by Camera 2 (lower)

(d)Vertical displacement time history monitored by Camera 2 (lower)

Fig. 4.28 Displacement monitoring results of the high formwork support system on 27 May

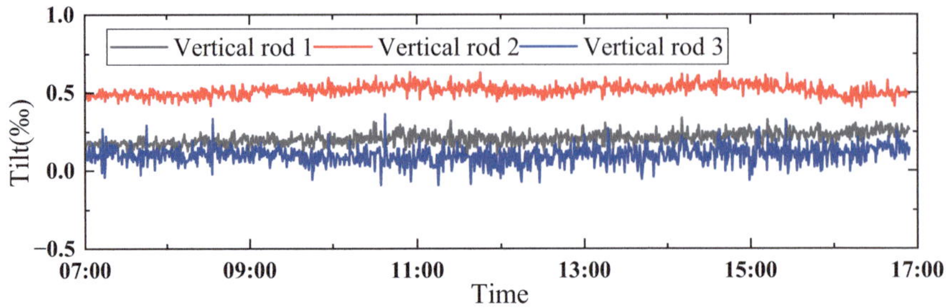

(e)Inclination time-history curve of vertical posts

Fig. 4.28 (continued)

(1) Illumination intensity: The M-DAVIM system achieves optimal accuracy within the illumination range of 200–400 lx. Measurement errors increase significantly under low or extreme illumination conditions but remain within engineering-acceptable tolerances.

Fog concentration: The integrated image restoration module based on generative adversarial networks effectively compensates for image degradation caused by fog, maintaining sub-millimeter accuracy even under dense fog conditions.

Camera elevation angle: The vision system maintains high accuracy within an elevation angle range of up to 30°.

Target size: When the target size is smaller than 18 pixels, measurement errors increase significantly; therefore, the marker size should be maintained above 18 pixels during outdoor deployment of the M-DAVIM system.

Camera vibration: The M-DAVIM system effectively suppresses vibration-induced errors in the frequency range of 5–50 Hz, achieving error reductions of up to 92% compared with traditional methods.

(2) Using the monitoring data from 27 May 2023 as an example, the full-day displacement and inclination of the high formwork support structure were analyzed. The time-history curves indicate that different vertical posts exhibit different horizontal and vertical displacements due to long-term construction effects; however, all monitored indicators remain below the alarm thresholds specified by the relevant code, confirming the structural safety. The inclination values are far below the inclination alarm limit, indicating that inclination is not the controlling monitoring factor in this project, and that greater emphasis should be placed on horizontal and vertical displacement monitoring.
(3) In this chapter, the monitoring alarm thresholds are fixed values determined according to the Technical Specification for Automated Monitoring of Steel Pipe Support Scaffolding in Construction (DBJ33/T1296-2023). However, in practical monitoring, structural conditions may continuously change due to factors such as component aging and load variation. Therefore, further research is required to develop scientifically sound methods for determining and dynamically updating alarm thresholds.

Chapter 5
Hybrid Data-Model-Driven Updating of Monitoring Alarm Thresholds and Short-Term Response Prediction for High Formwork Support Structures

Abstract A hybrid data-model-driven framework for monitoring threshold determination and dynamic updating is proposed, enabling short-term structural response prediction and construction load inversion (back-calculation). The framework consists of three core modules. The first module comprises a visual displacement monitoring system responsible for data acquisition, real-time warning, and continuous information transmission to the second module. The second module determines and updates monitoring thresholds using hybrid data-model-driven methods and provides extensive training samples for the third module. The third module employs the CNN-BiLSTM-AdaBoost algorithm to perform short-term prediction of structural displacement responses and inversion of construction loads, allowing displacement trends to be predicted up to one hour in advance and enabling early warning. The results demonstrate that the proposed monitoring threshold determination and updating method can be continuously refined in engineering practice. The buckling failure displacement predicted by the updated model exceeds the initial alarm threshold, indicating that the initial threshold is conservative and appropriately defined. Moreover, the CNN-BiLSTM-AdaBoost-based approach exhibits strong robustness and high prediction accuracy, enabling real-time response prediction and load inversion, and shows considerable potential for practical engineering monitoring applications.

Keywords High-formwork support system · Monitoring threshold update · Response prediction · CNN-BiLSTM-AdaBoost

For high formwork support structures, existing applications and studies lack scientific methods for determining and updating monitoring alarm thresholds. Moreover, current response alarm systems are triggered only when displacement exceeds preset limits and are unable to accurately predict response trends. To address these shortcomings, this study proposes a hybrid data-model-driven approach for determining and updating monitoring thresholds. Taking the actual high formwork project introduced in Chap. 4 as a case study, a closed-loop framework is established to achieve real-time

Q. Li et al., *A Hybrid Data-Model and AI-Driven Approach for Structural Monitoring in Hazardous Construction*, https://doi.org/10.1007/978-981-95-8688-2_5

perception of structural displacement and feedback-based updating, enabling continuous iterative updating of alarm thresholds. In addition, hybrid deep learning models and ensemble learning algorithms are employed to perform short-term prediction of structural displacement and rapid inversion of construction loads, thereby providing early warning for structural safety.

The analytical framework for updating monitoring alarm thresholds and short-term response prediction of high formwork support structures is shown in Figure 5.1, which mainly consists of three modules. The first module is the computer vision-based multi-point displacement monitoring system described in Chap. 4, which is used to perceive structural response data in real time, issue over-limit alarm signals, and provide continuous feedback to the second module. The second module adopts a hybrid data-model-driven approach to determine and update monitoring thresholds. The virtual-physical interaction process is as follows: first, a FEM is established by integrating the special design documentation for the high formwork system with the boundary conditions of the construction site; then, construction-stage design loads are applied to analyze the stress and deformation of the high formwork structure, and structural buckling failure analysis is performed to determine the initial alarm thresholds, which are used to guide on-site monitoring. As construction progresses, the structural state may change; therefore, accumulated measured data are used to periodically update the FEM. The updated model can more accurately reflect the actual stress state and deformation of the high formwork structure, allowing structural buckling failure analysis to be re-performed and alarm thresholds to be updated to continue guiding on-site monitoring. In addition, under extreme conditions where on-site monitoring may be infeasible due to safety or other constraints, data provided by the FEM can serve as a supplement to measured data, supplying sufficient training samples for the subsequent prediction module. The third module primarily employs the CNN-BiLSTM-AdaBoost algorithm for structural response prediction. This algorithm enables advanced prediction of displacement trends of the high formwork structure and issues early warning signals, facilitating timely and effective emergency measures.

5.1 Determination and Updating of Monitoring Alarm Thresholds

5.1.1 Initial Alarm Threshold Determination

(1) Model establishment

In this project, the high formwork support system adopts a cuplock-type steel pipe support frame. Adjustable screw jacks are used for both beam-bottom and slab-bottom vertical posts to enhance vertical load-bearing capacity. The material properties and cross-sectional information of the members are listed in Table 5.1.

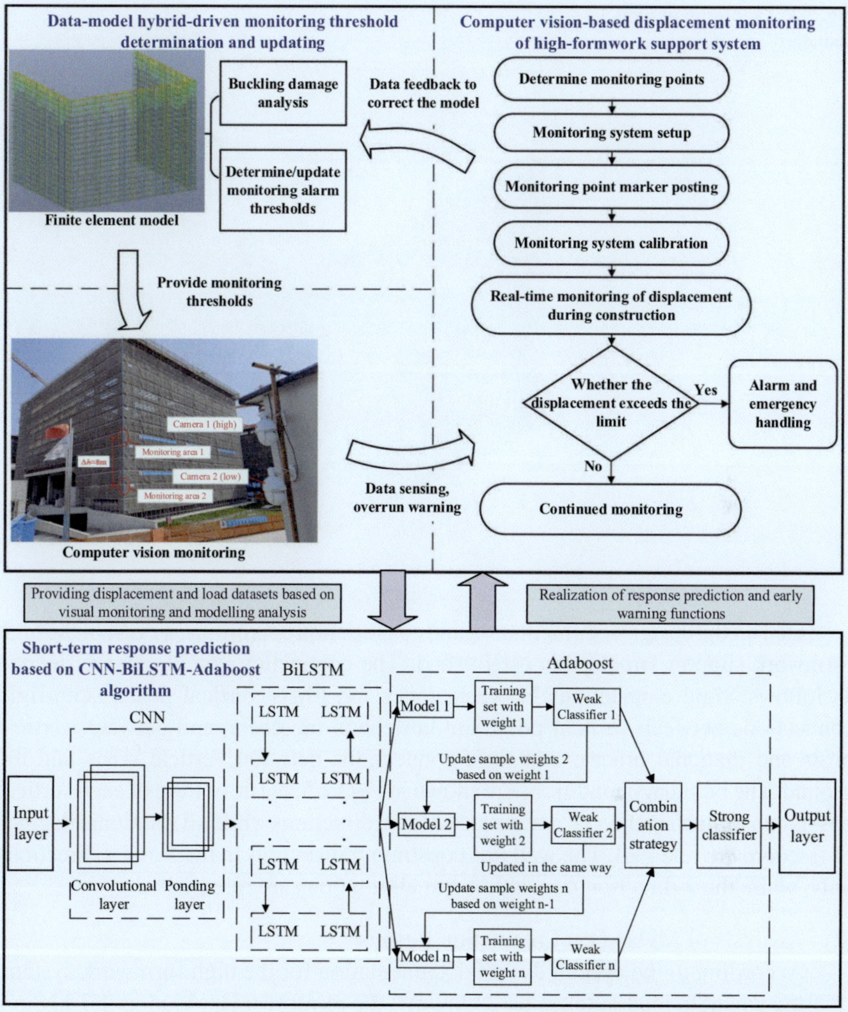

Fig. 5.1 Framework for updating monitoring alarm thresholds and short-term response prediction

Table 5.1 Material properties and cross-sectional information of members

Member	Yield strength (MPa)	Outer diameter (mm)	Wall thickness (mm)
Vertical rod	355	48.3	3.2
Cross rod	235	48.3	2.5
Tilt rod	235	33.7	2.3

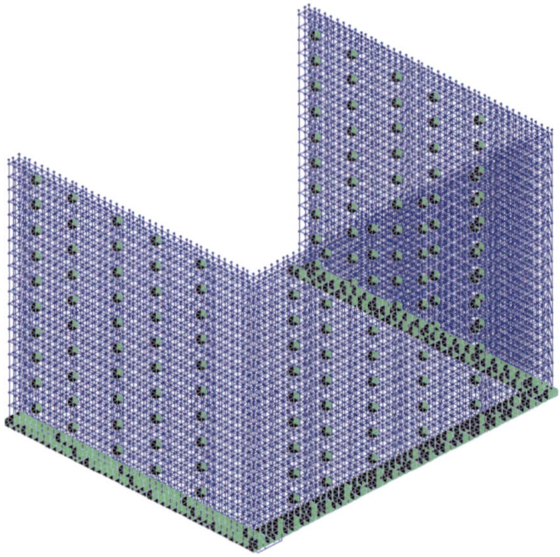

Fig. 5.2 Definition of model boundary conditions

Based on the design documentation and construction conditions, a FEM of the high formwork support structure is established. The connection conditions are defined as follows: rigid connections between upper and lower vertical posts; semi-rigid connections between vertical posts and horizontal members, and between vertical posts and diagonal braces; and rigid connections between vertical posts and the ground. The boundary conditions are shown in Figure 5.2. The bottom of each vertical post is constrained in the x, y, and z translational directions while all rotational degrees of freedom are released. The wall ties constrain translations in the x and y directions only, while the z-direction translation and all rotations are released.

(2) Analysis of Midas Gen Calculation Results

According to the special design documentation for the high formwork system, the concrete load is taken as 24 kN/m^3, the reinforcement load as 1.1 kN/m^3, the formwork load as 0.9 kN/m^3, the stacking load from construction personnel and equipment as 3 kN/m^2, and the standard value of horizontal load induced by pumping and pouring concrete as 0.08 kN/m^2. The calculated vertical load acting on the high formwork support system during construction is 4267.68 kN, and the horizontal load is 1725.84 kN. Assuming that the loads on all vertical posts are identical during construction, the high formwork support system consists of 408 vertical posts. Accordingly, each vertical post bears a vertical load of 10.46 kN and a horizontal load of 4.23 kN.

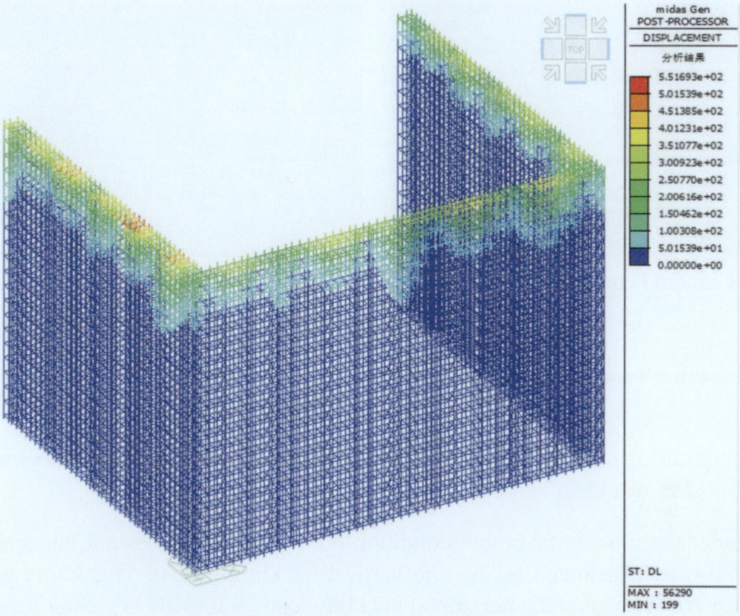

Fig. 5.3 Displacement contour plot obtained from Midas Gen

The FEM is analyzed using the buckling analysis command in the Midas Gen analysis module, yielding displacement contour plots and axial force diagrams, as well as buckling mode shapes. As shown in Figure 5.3, the maximum displacement of the high formwork support system is 5.52 mm, which is a very small deformation compared with the total height of the system (31,450 mm).

Considering both the finite element analysis results and on-site monitoring conditions, two elevation levels are selected for displacement monitoring to facilitate the presentation of the proposed monitoring alarm threshold updating framework, as shown in Figure 5.4. The monitoring points of Camera 1 are located at elevations of 15.4-16.9 m, where the finite element results indicate that the vertical displacement of the corner member (Member 2) is approximately 0.953-1.756 mm. The monitoring points of Camera 2 are located at an elevation of approximately 7.9 m, where the finite element results show that the vertical displacement of the corner member is about 2.29 mm. Similarly, the displacement and deformation of other members at different elevations can be obtained.

For high formwork support structures, buckling failure is one of the most common failure modes; therefore, buckling behavior must be considered in finite element analysis. Four working conditions are considered, accounting for whether the diagonal braces are effective and whether the loads are uniformly distributed. Taking a vertical load of 10.46 kN and a horizontal load of 4.23 kN as the initial input loads, the buckling eigenvalues and critical buckling loads of the formwork support system are calculated. By applying the buckling loads to the model, the critical displacements

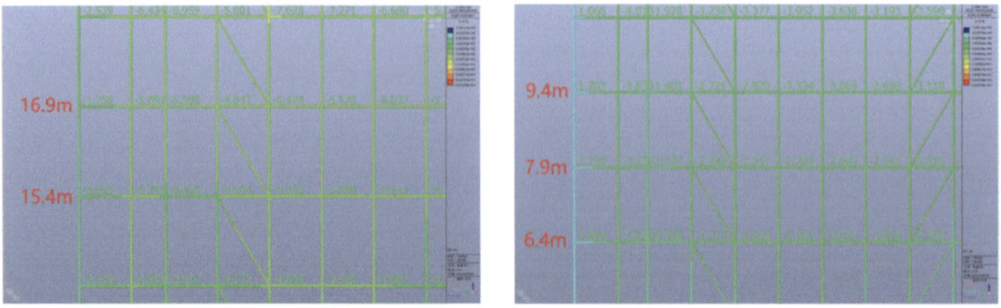

(a) Camera 1 (monitoring points at 15.4-16.9 m)

(b) Camera 2 (monitoring points at approximately 7.9 m)

Fig. 5.4 Settlement at monitoring points

of the monitoring nodes at the onset of buckling failure can be obtained, as listed in Table 5.2. The results indicate that:

(1) Under the most unfavorable condition (i.e., diagonal braces not engaged and uniformly distributed loads), the vertical buckling load of 18.2 kN is greater than 10.46 kN, and the horizontal buckling load of 7.36 kN is greater than 4.23 kN, indicating that the structure has a sufficient safety margin.

(2) At the onset of buckling failure, the displacements under uniformly distributed loads are smaller than those under non-uniformly distributed loads, indicating that larger displacements tend to activate the diagonal braces.

Table 5.2 Buckling failure analysis of the high formwork support structure

Working condition	Eigen value	Buckling load (kN)		Camera 1 displacement (mm, upper)		Camera 2 displacement (mm, lower)	
		Vertical	Horizontal	Vertical	Horizontal	Vertical	Horizontal
Diagonal braces active, uniform load	2.03	21.23	8.58	8.6	11.79	4.9	5.24
Diagonal braces inactive, uniform load	1.74	18.20	7.36	7.5	9.91	4.3	3.9
Diagonal braces active, non-uniform load	5.72	59.83	24.20	15.88	13.83	7.8	7.1
Diagonal braces inactive, non-uniform load	5.40	56.48	22.84	15.18	12.9	7.5	6.5

(3) Based on the most unfavorable condition, the monitoring alarm thresholds for the high formwork support structure are set as follows: for Camera 1, 7.5 mm in the vertical direction and 9.91 mm in the horizontal direction; for Camera 2, 4.3 mm in the vertical direction and 3.9 mm in the horizontal direction. All these values are lower than the upper limits specified in the Technical Specification for Automated Monitoring of Steel Pipe Support Scaffolding in Construction (DBJ33/T1296-2023), namely 8 mm for vertical settlement and 12 mm for horizontal displacement, indicating that the selected thresholds are reasonable.

5.1.2 Updating of Alarm Thresholds

To demonstrate the process of updating alarm thresholds, twelve sets of maximum vertical and horizontal displacement data were selected from six monitoring points during the concrete pouring stage and compared with the finite element calculation results, as shown in Table 5.3. Among these, the relative errors between the finite element results and the measured displacements for seven data sets are controlled within 20%, while the remaining five data sets exhibit relatively large errors. These include the vertical displacements of Member 1 (upper), Member 2 (upper), and Member 3 (upper), as well as the horizontal displacements of Member 3 (upper) and Member 1 (lower). The possible reasons for the larger errors in these five data sets are as follows: (1) the quality of node connections at these monitoring points during construction may be inadequate; (2) the cuplock support system has relatively high strength and stiffness, resulting in small actual displacements, for which small absolute errors can lead to large relative errors; and (3) the vertical posts may have developed local damage after repeated reuse, causing the initial FEM to inadequately represent the actual structural behavior.

(1) FEM Updating Method

To more accurately reflect the actual stress state and deformation behavior of the structure, it is necessary to periodically update the FEM. The first step is to perform a sensitivity analysis to evaluate the influence of different parameters

Table 5.3 Comparison between finite element calculated displacements (before updating) and measured displacements

Measurement Point	Vertical Direction (mm)			Horizontal Direction (mm)		
	FEM	Vision	Error (%)	FEM	Vision	Error (%)
Member 1 (upper)	1.65	0.9	83.3	3.93	4.3	8.6
Member 2 (upper)	1.40	1.9	26.3	3.93	3.6	9.2
Member 3 (upper)	5.50	3.6	52.8	5.52	4.2	31.4
Member 1 (lower)	3.32	3.6	7.8	4.24	3.1	36.8
Member 2 (lower)	2.26	1.9	18.9	4.25	3.8	11.8
Member 3 (lower)	3.51	3.8	7.6	2.18	2.5	12.8

on the model, so as to identify appropriate updating parameters. The sensitivity calculation formula is given in Eq. (3.2). In this study, the candidate parameters include joint stiffness, member density, elastic modulus, and member wall thickness. The results of the parameter sensitivity analysis are summarized in Table 5.4. When the variation rate of all parameters is set to 10%, the elastic modulus and wall thickness of the vertical posts exhibit the greatest influence on the displacement of the high formwork support structure and are therefore selected as the updating parameters.

By comparing the root mean square error (RMSE) between the finite element calculation results and the field-measured data before and after model updating, the effectiveness of the FEM updating can be evaluated. Figure 5.5 shows the comparison of displacement errors between the finite element results and measured data before and after updating. After updating, 11 out of the 12 data sets have relative errors between the finite element results and measured values controlled within 20%, and the RMSE decreases from 0.784 mm to 0.132 mm. This indicates that the updated model can more accurately reflect the actual stress state and deformation behavior of the structure.

Table 5.4 Results of parameter sensitivity analysis

Parameter (10% variation)	Displacement change rate	Sensitivity
Joint stiffness	−16.80	1.68
Member density	−17.20	1.72
Elastic modulus of vertical posts	−29.50	2.95
Elastic modulus of horizontal members	−10.70	1.07
Wall thickness of vertical posts	−27.10	2.71
Wall thickness of horizontal members	−15.50	1.55

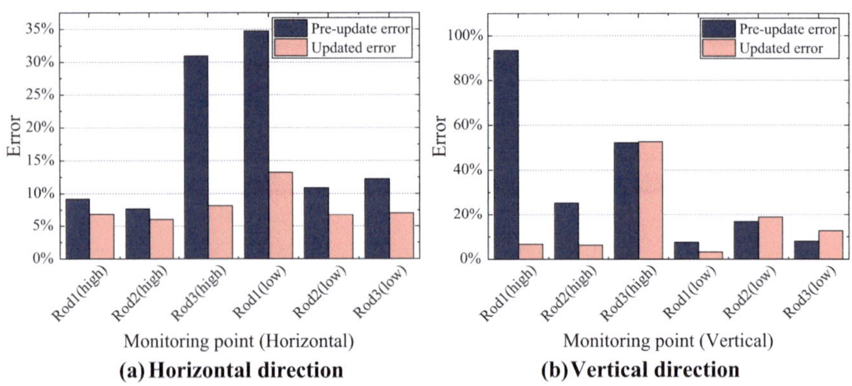

Fig. 5.5 Comparison of displacement errors between FEM and measured data before and after updating

(2) Rationality Analysis of Alarm Thresholds
Buckling failure analysis is re-performed using the updated FEM to determine the critical displacements, based on which the rationality of the initial alarm thresholds is evaluated. If the alarm thresholds are set too high, structural failure may occur without triggering an alarm, compromising construction safety. Conversely, if the alarm thresholds are set too low, frequent false alarms may occur, adversely affecting construction progress and consuming emergency response resources. Therefore, the alarm thresholds before updating should be slightly lower than the buckling failure displacements obtained after updating to ensure rationality. Figure 5.6 shows the long-term displacement monitoring results and a comparison of alarm thresholds before and after updating. It can be observed that, with time accumulation, both the horizontal drift and vertical settlement of the high formwork support structure gradually increase. Meanwhile, the updated critical horizontal displacements for Camera 1 and Camera 2 are 11.2 mm (> 9.91 mm) and 7.95 mm (> 7.5 mm), respectively, and the updated critical vertical displacements for Camera 1 and Camera 2 are 5.16 mm (> 3.9 mm) and 4.51 mm (> 4.3 mm), respectively. All updated alarm thresholds exceed the initial ones, indicating that the initial alarm thresholds were conservatively set and can effectively prevent structural failure before reaching the preset alarm limits.

5.2 Machine Learning-Based Response Prediction of High Formwork Support Structures

5.2.1 Short-Term Structural Displacement Prediction

To establish a short-term structural displacement prediction model based on the CNN-BiLSTM-AdaBoost algorithm, this section uses computer vision-based monitoring results as the primary data source, supplemented by displacement data simulated from the FEM. Working days are divided into hourly time intervals, and the maximum displacement within each interval is selected. The peak displacement of the current interval is used as the model input, while the peak displacement of the subsequent interval is used as the model output to construct the dataset. The updated critical displacement values of the model are set as the early warning thresholds, enabling prediction of structural instability and damage risk one hour in advance.

A total of 590 data samples are selected from the monitoring data and divided into a training set and a testing set. The detailed parameters of the training and testing sets are listed in Table 5.5.

Figures 5.7a–d and Table 5.6 present the displacement prediction results for Member 1 (Camera 1). A comparison is conducted among the CNN-BiLSTM-AdaBoost, CNN, and BiLSTM models. The results indicate that the CNN-BiLSTM-AdaBoost model produces predictions that are closest to the measured values, with

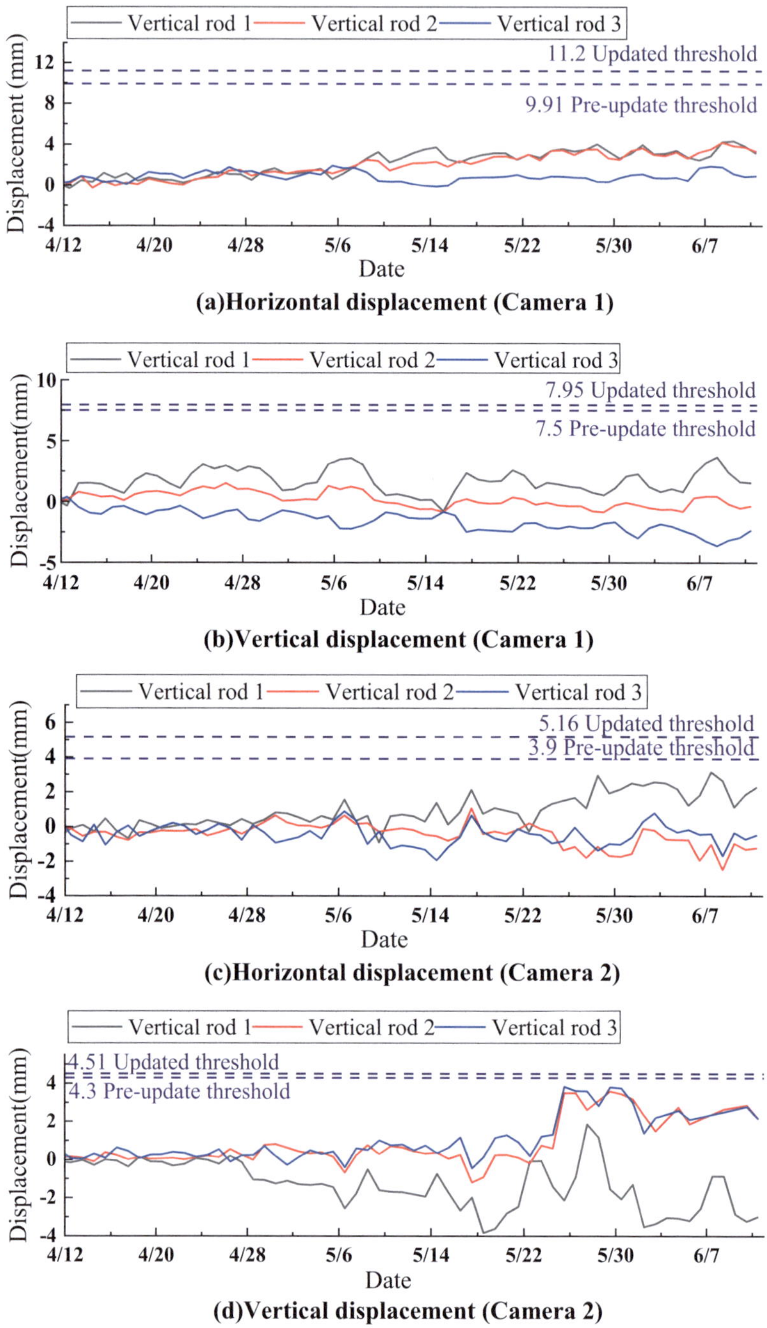

Fig. 5.6 Comparison of alarm thresholds before and after updating

Table 5.5 Parameters of the training and testing sets for displacement prediction

Monitoring Point	Direction	Original sampling frequency (Hz)	Adjusted sampling frequency (Hz)	Training samples	Testing samples
Member 1 (Camera 1)	Horizontal	1/30	1/3600	456	134
	Vertical			420	170
Member 1 (Camera 2)	Horizontal	1/30	1/3600	444	146
	Vertical			420	170
Member 2 (Camera 1)	Horizontal	1/30	1/3600	420	170
	Vertical			420	170
Member 2 (Camera 2)	Horizontal	1/30	1/3600	420	170
	Vertical			420	170
Member 3 (Camera 1)	Horizontal	1/30	1/3600	450	140
	Vertical			444	146
Member 3 (Camera 2)	Horizontal	1/30	1/3600	450	140
	Vertical			444	146

smaller and more stable errors, demonstrating superior short-term prediction capability and robustness. Figures 5.7e–h show the regression plots of the CNN-BiLSTM-AdaBoost model for the horizontal and vertical directions, which exhibit strong correlations between the predicted and measured values. To further compare prediction accuracy, the error metrics of each model are calculated and summarized in Table 5.6. As shown in the table, the CNN-BiLSTM-AdaBoost model achieves the smallest RMSE and maximum error, as well as the R^2 values closest to 1, for both horizontal and vertical displacement predictions, indicating its superior predictive performance.

Figures 5.8a-d and Table 5.7 present the displacement prediction results for Member 1 (Camera 2). Figures 5.9 and Table 5.8 show the prediction results of the displacement response for Member 2 (Camera 1). Figures 5.10 and Table 5.9 present the prediction results for Member 2 (Camera 2). Figures 5.11 and Table 5.10 show the prediction results for Member 3 (Camera 1), while Figure 5.12 and Table 5.11 present the prediction results for Member 3 (Camera 2). The three algorithms are compared in all cases. The results indicate that the CNN-BiLSTM-AdaBoost model produces predictions closest to the measured values, with relatively smaller errors. The regression plots for both horizontal and vertical directions exhibit strong correlations between predicted and measured values. From the error metrics summarized in the tables, it can be observed that the CNN-BiLSTM-AdaBoost model achieves the smallest RMSE and maximum error, and the R^2 values closest to 1 for both horizontal and vertical displacement predictions, demonstrating superior predictive performance.

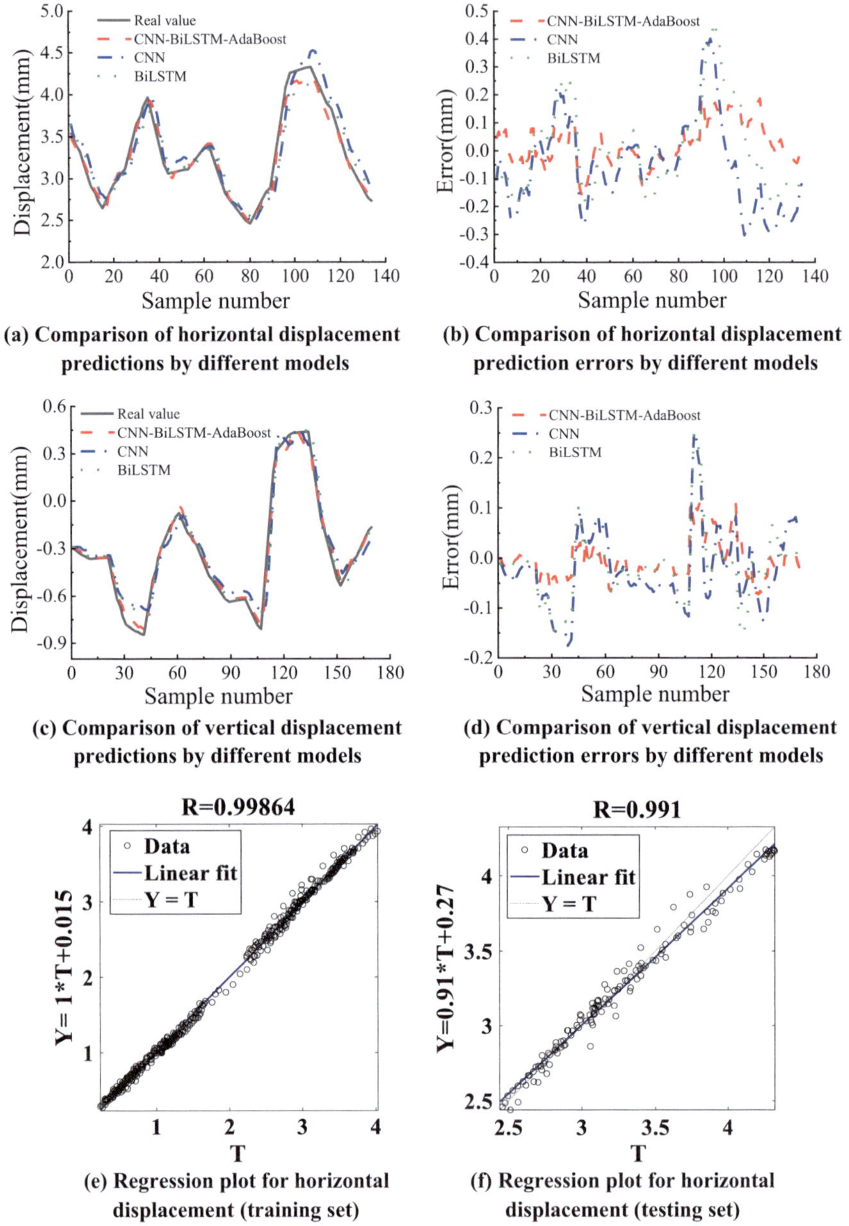

Fig. 5.7 Prediction results of displacement responses for Member 1 (Camera 1)

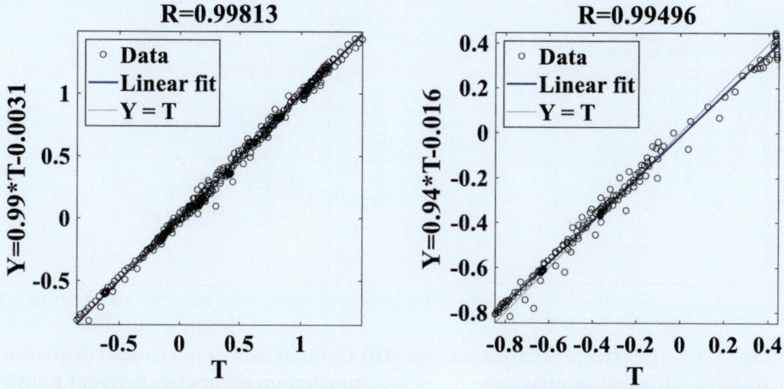

(g) Regression plot for vertical displacement (training set)

(h) Regression plot for vertical displacement (testing set)

Fig. 5.7 (continued)

Table 5.6 Comparison of prediction performance metrics of different models for Member 1 (Camera 1)

Direction	Metric	CNN	BiLSTM	CNN-BiLSTM-AdaBoost
Horizontal	*RMSE*/mm	0.1683	0.1530	0.0798
	R^2	0.9370	0.9107	0.9821
	Max error/mm	0.401	0.463	0.195
Vertical	*RMSE*/mm	0.0793	0.0718	0.0403
	R^2	0.9568	0.9613	0.9899
	Max error/mm	0.252	0.248	0.113

In summary, the CNN-BiLSTM-AdaBoost model performs excellently in predicting both horizontal and vertical displacements, exhibiting good fitting capability and high prediction accuracy.

5.2.2 Rapid Inversion of Upper Construction Loads

In current engineering practice, effective means for real-time monitoring of upper construction loads acting on high formwork support structures are lacking. Existing approaches often rely on manual records and estimations by construction personnel, which cannot eliminate potential adverse load hazards during construction. In this section, vision-based displacement monitoring data, manually recorded construction loads, and model-based analytical load and displacement data are integrated. The displacement data at monitoring points are used as inputs, while the corresponding loads are used as outputs to construct training and testing datasets for load inversion.

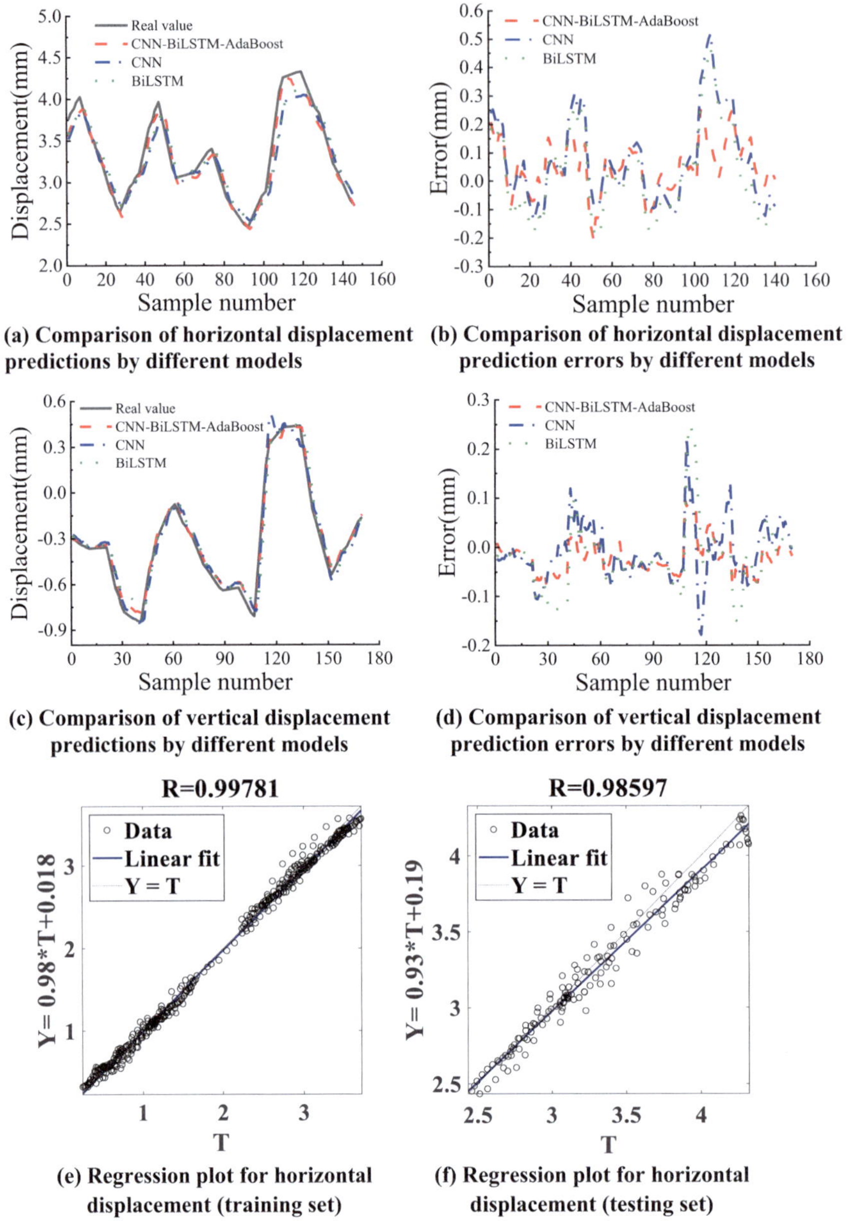

(a) Comparison of horizontal displacement predictions by different models

(b) Comparison of horizontal displacement prediction errors by different models

(c) Comparison of vertical displacement predictions by different models

(d) Comparison of vertical displacement prediction errors by different models

(e) Regression plot for horizontal displacement (training set)

(f) Regression plot for horizontal displacement (testing set)

Fig. 5.8 Prediction results of displacement responses for Member 1 (Camera 2)

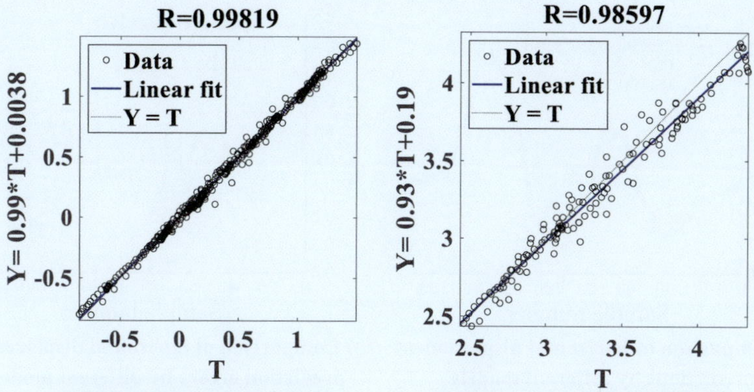

(g) Regression plot for vertical displacement (training set)

(h) Regression plot for vertical displacement (testing set)

Fig. 5.8 (continued)

Table 5.7 Comparison of prediction performance metrics of different models for Member 1 (Camera 2)

Direction	Metric	CNN	BiLSTM	CNN-BiLSTM-AdaBoost
Horizontal	*RMSE*/mm	0.166	0.155	0.100
	R^2	0.897	0.908	0.972
	Max error/mm	0.520	0.469	0.251
Vertical	*RMSE*/mm	0.117	0.072	0.041
	R^2	0.958	0.957	0.972
	Max error/mm	0.225	0.250	0.096

The training set consists of 560 samples, and the testing set consists of 230 samples. Consistent with the short-term displacement prediction described in Section 5.2.1, the sampling interval is set to 1 hour.

Figures 5.13a–d compare the performance of different models in predicting horizontal and vertical loads. As shown in the figures, the CNN-BiLSTM-AdaBoost model produces predictions closest to the measured values, with a relatively uniform error distribution, indicating greater stability and reliability. As illustrated in Figures 5.13e–h, the regression analysis further demonstrates that, for both horizontal and vertical load predictions, the CNN-BiLSTM-AdaBoost model achieves R^2 values close to 1, indicating strong correlations between predicted and actual loads. Table 5.12 summarizes the error metrics of the different models. For horizontal load prediction, the CNN-BiLSTM-AdaBoost model exhibits the lowest RMSE, the highest R^2, and the smallest errors, demonstrating high accuracy and stability. For vertical load prediction, although the BiLSTM model achieves the highest R^2, it also

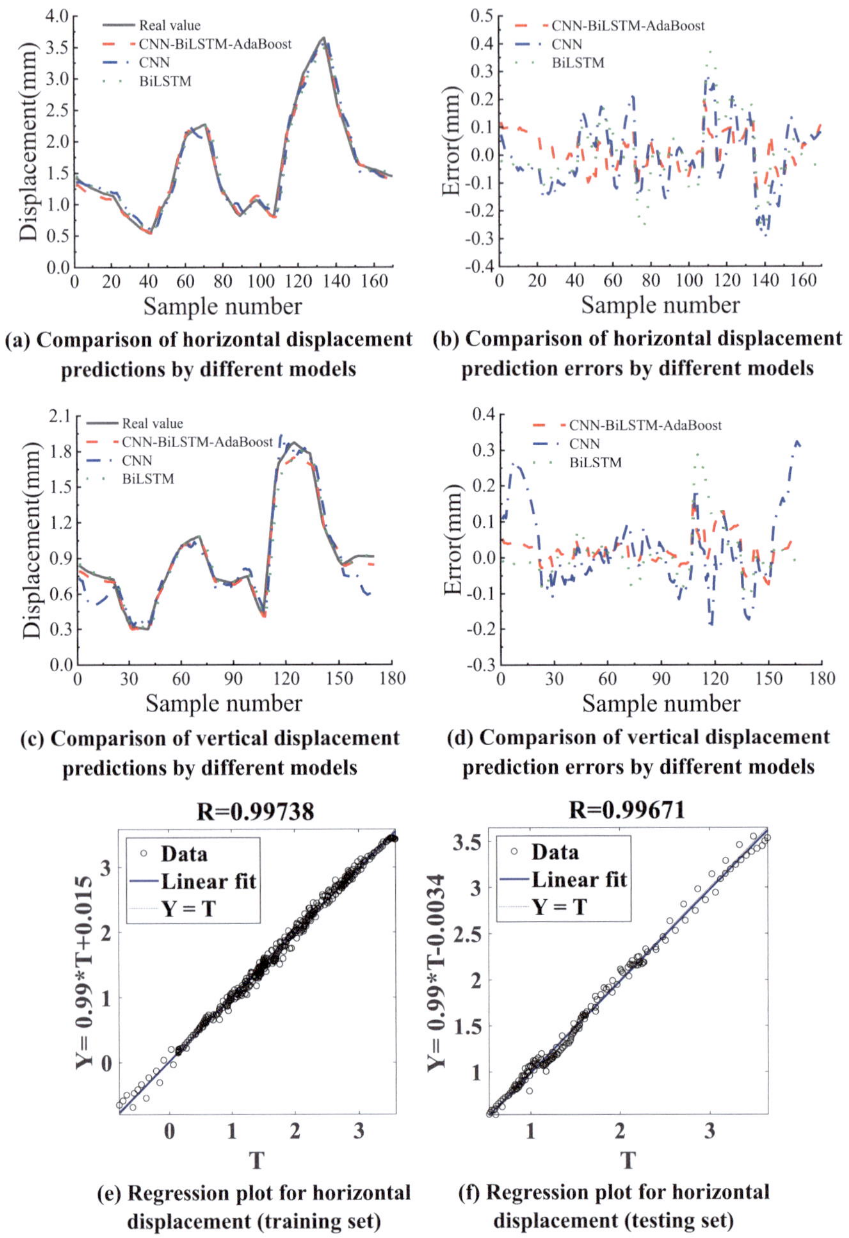

(a) Comparison of horizontal displacement predictions by different models
(b) Comparison of horizontal displacement prediction errors by different models
(c) Comparison of vertical displacement predictions by different models
(d) Comparison of vertical displacement prediction errors by different models
(e) Regression plot for horizontal displacement (training set)
(f) Regression plot for horizontal displacement (testing set)

Fig. 5.9 Prediction results of displacement responses for Member 2 (Camera 1)

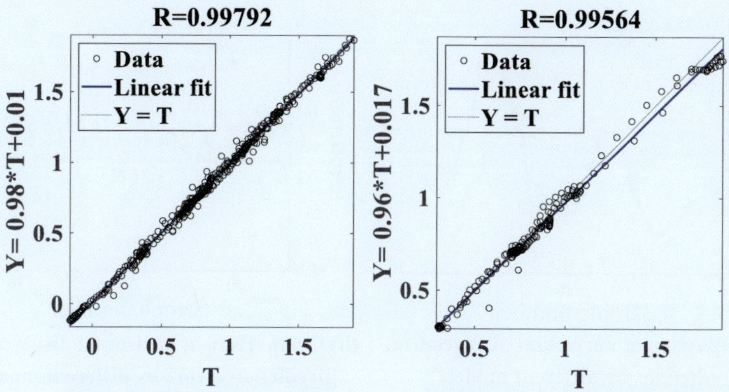

(g) Regression plot for vertical displacement (training set)

(h) Regression plot for vertical displacement (testing set)

Fig. 5.9 (continued)

Table 5.8 Comparison of prediction performance metrics of different models for Member 2 (Camera 1)

Direction	Metric	CNN	BiLSTM	CNN-BiLSTM-AdaBoost
Horizontal	*RMSE*/mm	0.113	0.121	0.065
	R^2	0.979	0.976	0.993
	Max error/mm	0.299	0.371	0.193
Vertical	*RMSE*/mm	0.168	0.074	0.046
	R^2	0.892	0.968	0.991
	Max error/mm	0.324	0.289	0.178

shows the largest RMSE and maximum error, indicating relatively inferior prediction accuracy compared with the other two models. In contrast, the CNN-BiLSTM-AdaBoost model achieves a lower RMSE and smaller errors, confirming its reliable predictive performance.

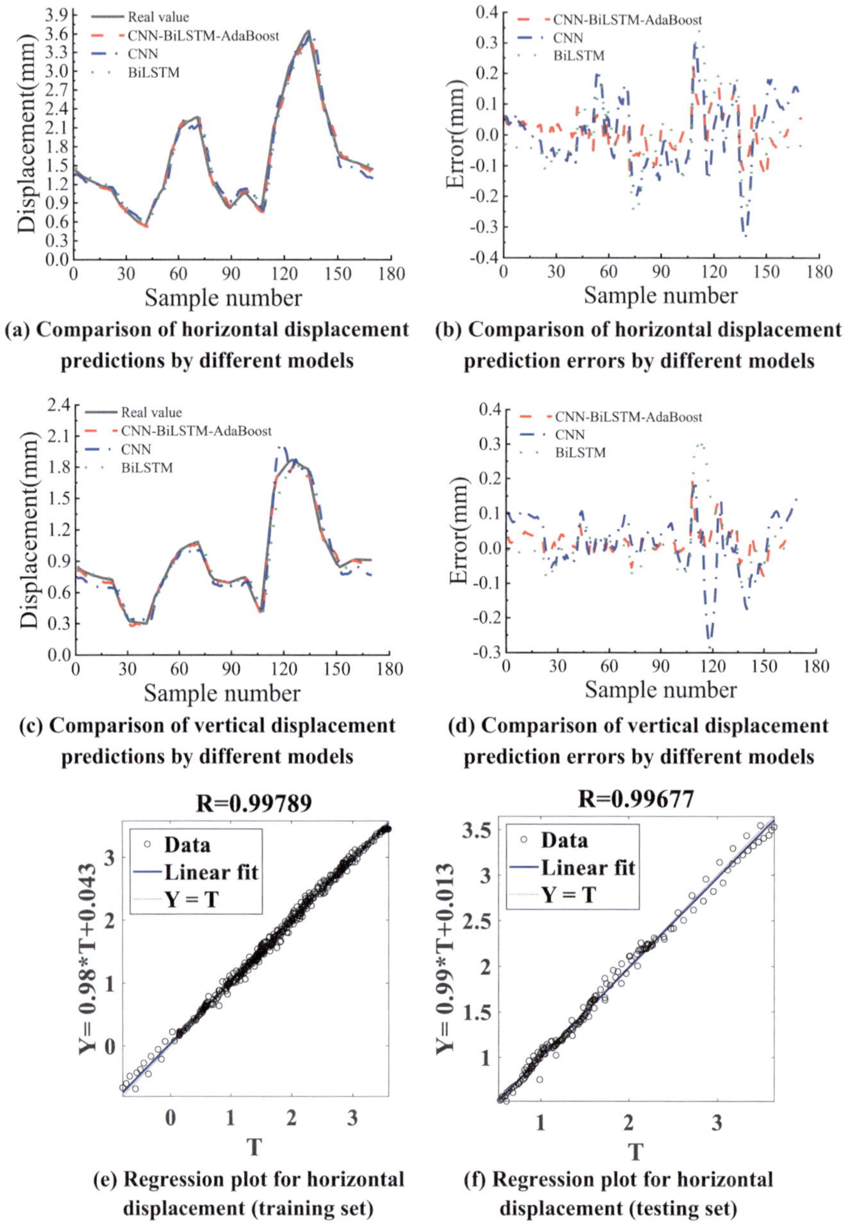

Fig. 5.10 Prediction results of displacement responses for Member 2 (Camera 2)

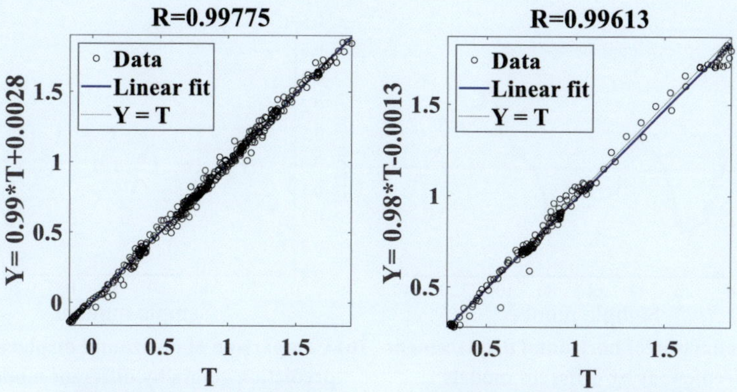

(g) Regression plot for vertical displacement (training set)

(h) Regression plot for vertical displacement (testing set)

Fig. 5.10 (continued)

Table 5.9 Comparison of prediction performance metrics of different models for Member 2 (Camera 2)

Direction	Metric	CNN	BiLSTM	CNN-BiLSTM-AdaBoost
Horizontal	*RMSE*/mm	0.114	0.113	0.064
	R^2	0.979	0.979	0.994
	Max error/mm	0.338	0.346	0.231
Vertical	*RMSE*/mm	0.082	0.079	0.040
	R^2	0.960	0.963	0.992
	Max error/mm	0.288	0.317	0.194

5.3 Summary

This chapter proposes a hybrid data-model-driven framework for determining and updating monitoring thresholds of high formwork support systems and for performing short-term response prediction. First, a FEM is established to determine the initial alarm thresholds. Then, a vision-based structural displacement monitoring system is employed to achieve real-time perception of structural responses and real-time data feedback, providing a foundation for continuous updating of monitoring thresholds. Finally, the CNN-BiLSTM-AdaBoost algorithm is adopted to perform short-term prediction of structural displacement and rapid inversion of construction loads, thereby providing early warning for high formwork support structures. The main conclusions are summarized as follows:

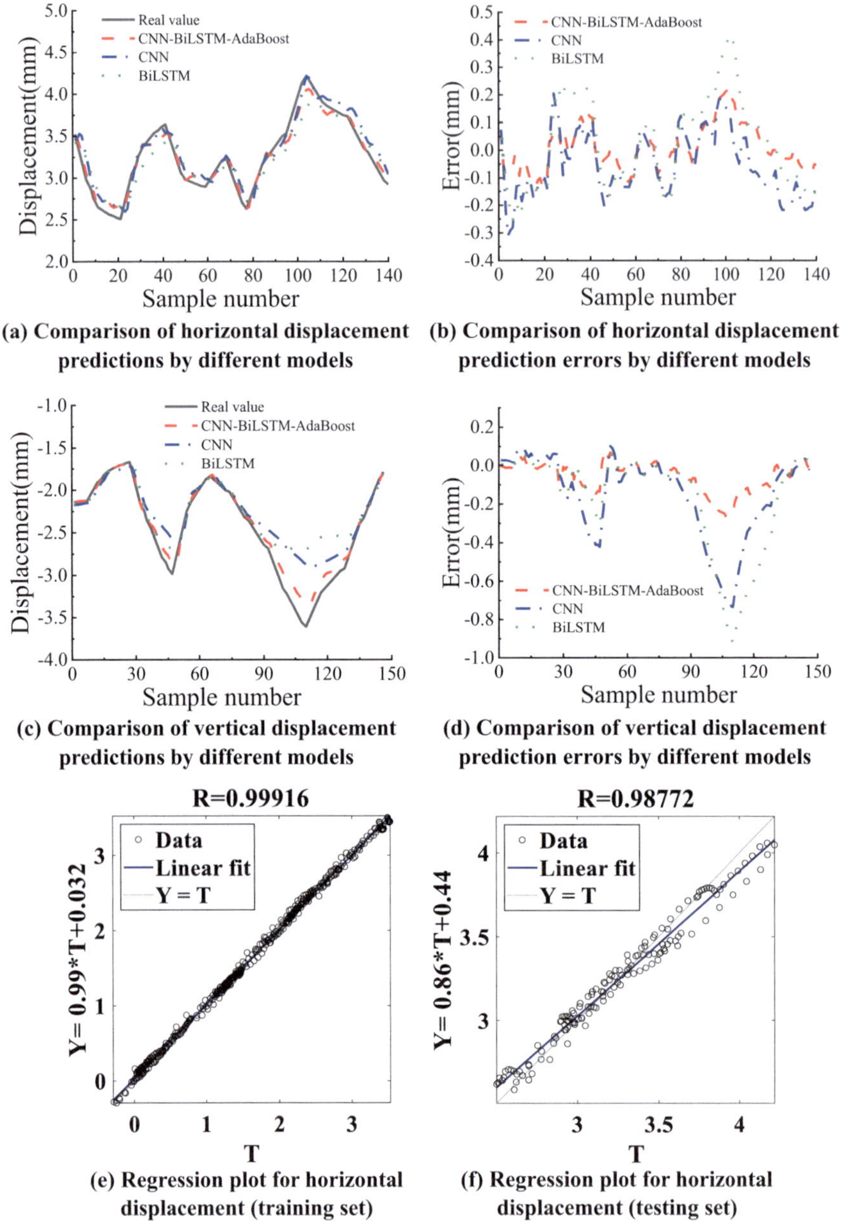

(a) Comparison of horizontal displacement predictions by different models

(b) Comparison of horizontal displacement prediction errors by different models

(c) Comparison of vertical displacement predictions by different models

(d) Comparison of vertical displacement prediction errors by different models

(e) Regression plot for horizontal displacement (training set)

(f) Regression plot for horizontal displacement (testing set)

Fig. 5.11 Prediction results of displacement responses for Member 3 (Camera 1)

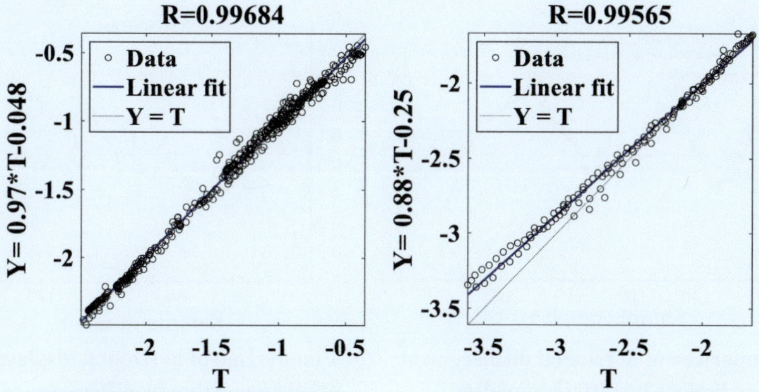

(g) Regression plot for vertical displacement (training set)

(h) Regression plot for vertical displacement (testing set)

Fig. 5.11 (continued)

Table 5.10 Comparison of prediction performance metrics of different models for Member 3 (Camera 1)

Direction	Metric	CNN	BiLSTM	CNN-BiLSTM-AdaBoost
Horizontal	*RMSE*/mm	0.127	0.167	0.083
	R^2	0.914	0.849	0.976
	Max error/mm	0.309	0.417	0.281
Vertical	*RMSE*/mm	0.266	0.315	0.097
	R^2	0.775	0.691	0.991
	Max error/mm	0.736	0.922	0.281

(1) The monitoring alarm threshold determination and updating framework proposed in this chapter is a dynamic and continuously improving process. The alarm thresholds must be iteratively updated to accurately match the actual structural characteristics. The updated buckling failure displacements of the structure are all greater than the initial alarm thresholds, indicating that the initial thresholds were conservatively set and that the selected values are reasonable.

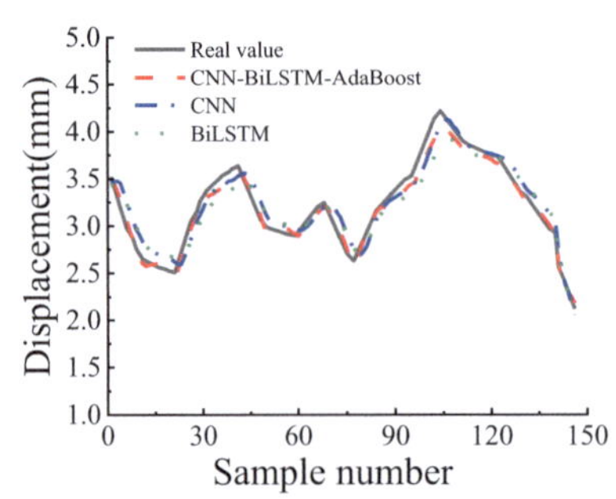

(a) Comparison of horizontal displacement predictions by different models

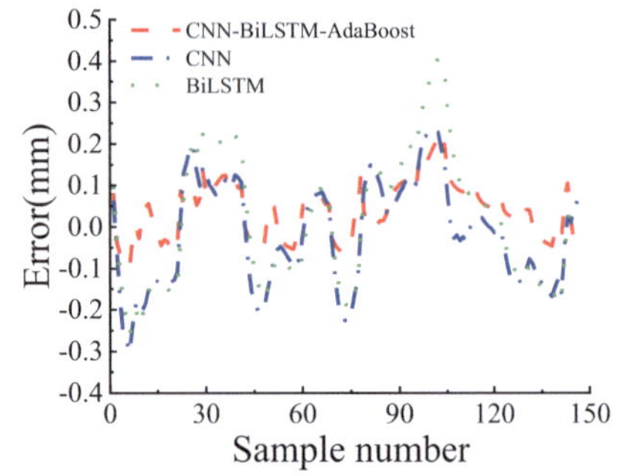

(b) Comparison of horizontal displacement prediction errors by different models

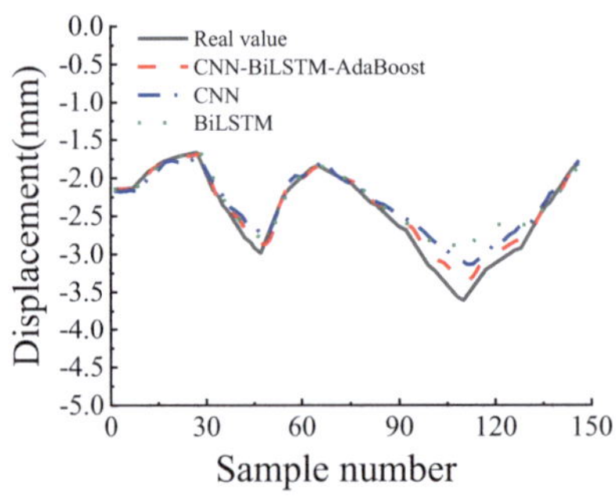

(c) Comparison of vertical displacement predictions by different models

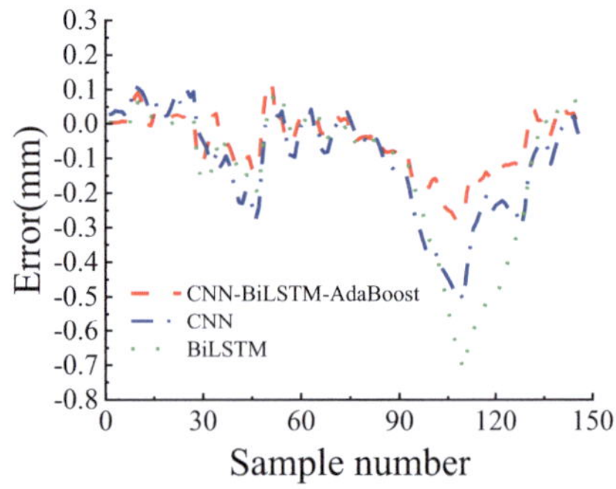

(d) Comparison of vertical displacement prediction errors by different models

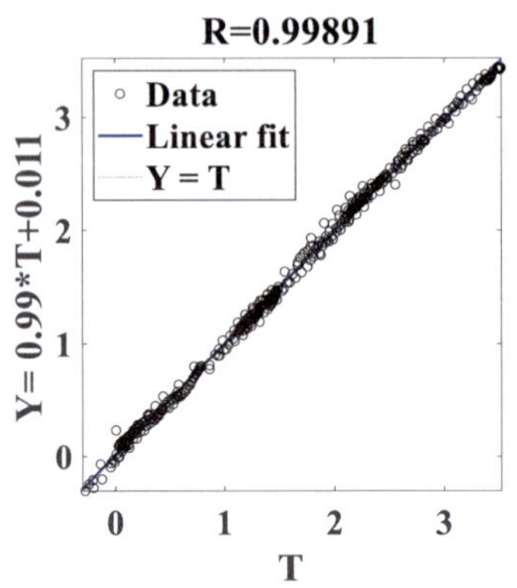

(e) Regression plot for horizontal displacement (training set)

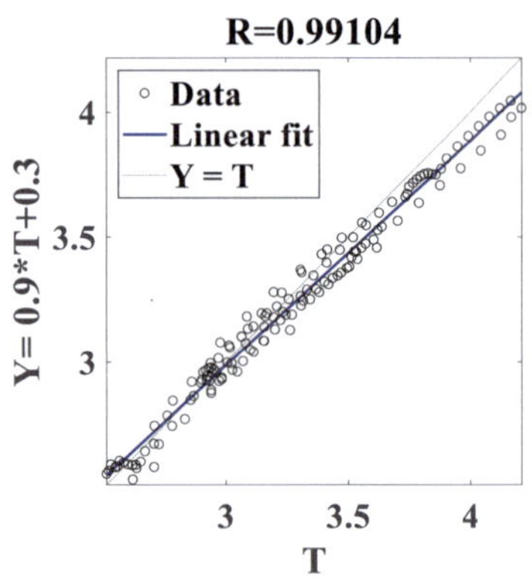

(f) Regression plot for horizontal displacement (testing set)

Fig. 5.12 Prediction results of displacement responses for Member 3 (Camera 2)

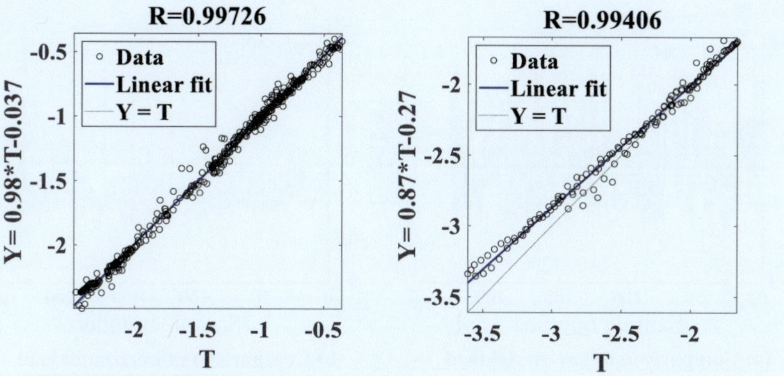

(g) Regression plot for vertical displacement (training set)

(h) Regression plot for vertical displacement (testing set)

Fig. 5.12 (continued)

Table 5.11 Comparison of prediction performance metrics of different models for Member 3 (Camera 2)

Direction	Metric	CNN	BiLSTM	CNN-BiLSTM-AdaBoost
Horizontal	*RMSE*/mm	0.130	0.164	0.078
	R^2	0.907	0.853	0.982
	Max error/mm	0.301	0.408	0.221
Vertical	*RMSE*/mm	0.189	0.245	0.103
	R^2	0.882	0.804	0.988
	Max error/mm	0.507	0.715	0.287

(2) The CNN-BiLSTM-AdaBoost hybrid deep learning model demonstrates strong robustness and high prediction accuracy. This algorithm is capable of short-term prediction of structural responses and real-time inversion of construction loads, highlighting its significant potential for practical application in engineering monitoring.

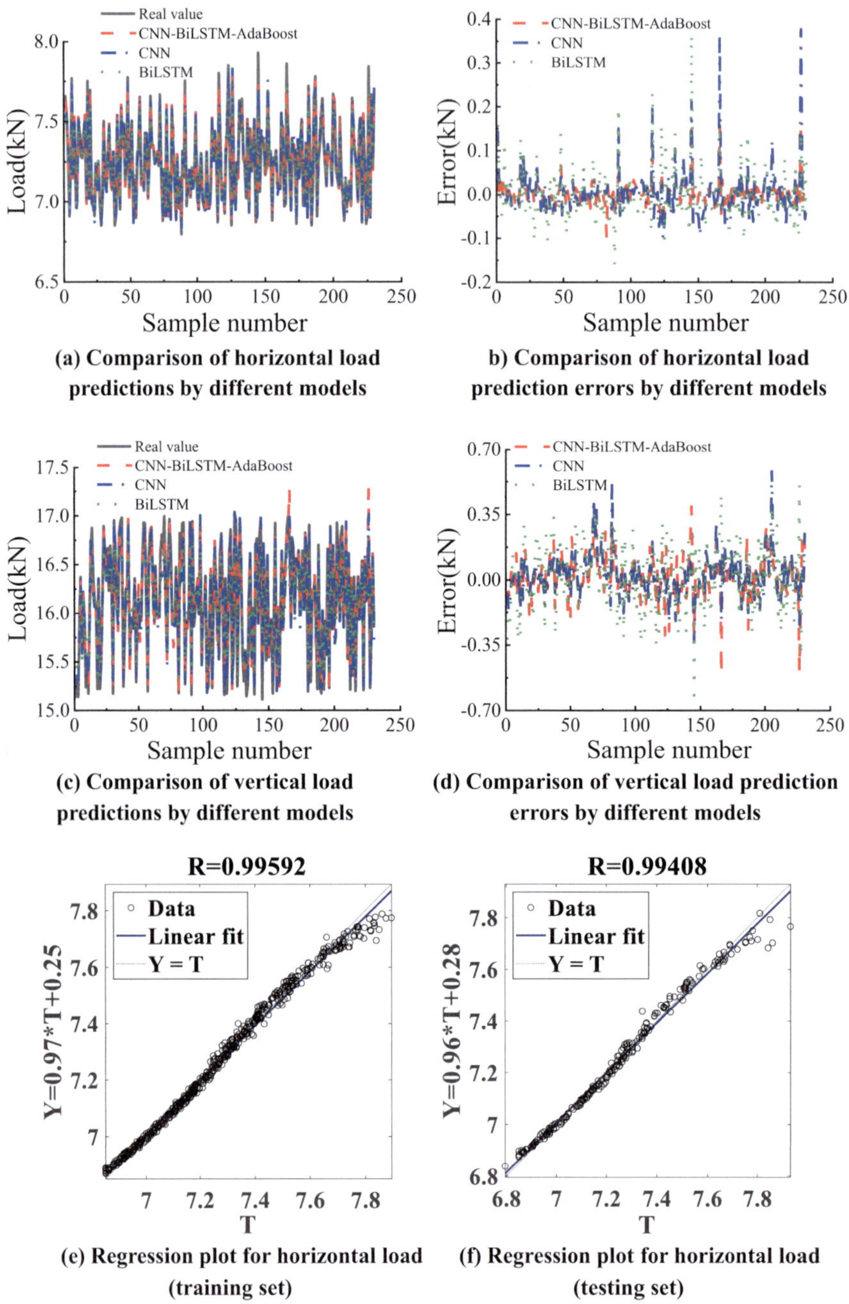

(a) Comparison of horizontal load predictions by different models

b) Comparison of horizontal load prediction errors by different models

(c) Comparison of vertical load predictions by different models

(d) Comparison of vertical load prediction errors by different models

(e) Regression plot for horizontal load (training set)

(f) Regression plot for horizontal load (testing set)

Fig. 5.13 Load inversion prediction results

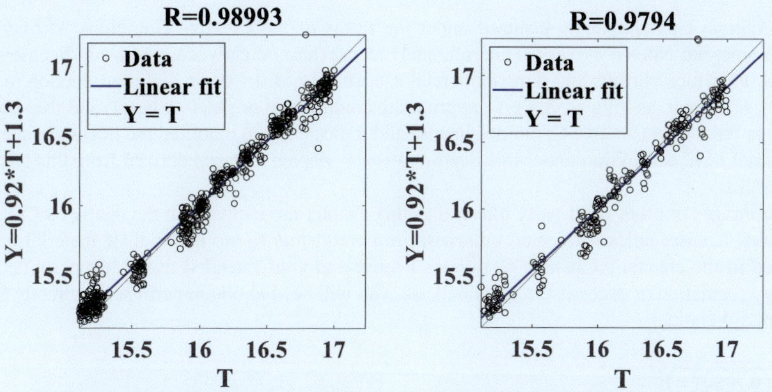

(g) Regression plot for vertical load (training set)

(h) Regression plot for vertical load (testing set)

Fig. 5.13 (continued)

Table 5.12 Comparison of prediction performance metrics of different models

Direction	Metric	CNN	BiLSTM	CNN-BiLSTM-AdaBoost
Horizontal	*RMSE*/mm	0.0464	0.0615	0.0289
	R^2	0.9572	0.9807	0.9882
	Max error/mm	0.377	0.367	0.164
Vertical	*RMSE*/mm	0.1175	0.1850	0.1182
	R^2	0.9617	0.9657	0.9592
	Max error/mm	0.583	0.620	0.510

Chapter 6
Conclusions and Future Perspectives

6.1 Key Innovations and Contributions

This study adopts a hybrid data-model-driven approach with a focus on structural safety monitoring and early warning for hazard-prone construction projects. For tower cranes, research was conducted on Internet of Things (IoT)-based monitoring, structural dynamic response analysis, and response prediction for early warning. For high-formwork support systems, studies were carried out on computer vision-based monitoring, dynamic updating of monitoring and warning thresholds, and short-term response prediction. The main innovations are summarized as follows:

(1) Tower crane structures

A hybrid data-model-driven dynamic response prediction framework for tower cranes under extreme typhoon conditions is proposed. This framework effectively addresses the shortage of extreme-event samples in measured data and significantly improves the accuracy and reliability of predicting "gray swan" typhoon events.

Furthermore, a wind-induced dynamic response prediction method for tower cranes based on a CNN-BiLSTM-AdaBoost ML model is developed, enabling accurate capture of response trends and providing early warning capability.

(2) High-formwork support structures

A hybrid data-model-driven framework for determining and dynamically updating monitoring and warning thresholds of high-formwork support structures is proposed, allowing continuous iteration of alarm thresholds and improving early warning accuracy.

A visual displacement monitoring system for high-formwork structures is established, enabling real-time perception of structural responses and timely data feedback, which lays a solid foundation for warning threshold updating.

In addition, a CNN-BiLSTM-AdaBoost-based method for short-term displacement prediction and rapid inversion of upper construction loads is proposed, providing effective early warning for high-formwork support structures。

Q. Li et al., *A Hybrid Data-Model and AI-Driven Approach for Structural Monitoring in Hazardous Construction*, https://doi.org/10.1007/978-981-95-8688-2_6

6.2 Main Conclusions

This study proposes a hybrid data-model-driven method to predict tower crane displacement responses under extreme typhoon conditions and to realize the updating of monitoring and warning thresholds as well as short-term response prediction for high-formwork support structures. In the tower crane study, an actual construction project tower crane was taken as an example, and an IoT-based real-time displacement monitoring system was developed to obtain load-response data of the tower crane during construction. Meanwhile, a FEM was established to simulate tower crane displacements under various extreme wind loads, compensating for the lack of extreme-event samples in the monitoring data. Then, based on the CNN-BiLSTM-AdaBoost algorithm, both data-driven and hybrid data-model-driven methods were adopted to predict the top displacement of the tower crane under normal wind and typhoon conditions. In the high-formwork structure study, the influences of different external factors on the visual system were first investigated. Then, taking an actual high-formwork structure as an example, a visual monitoring system and a FEM were established, forming a closed-loop framework for real-time perception and feedback updating of structural responses, which enables the determination and continuous iterative updating of alarm thresholds. Finally, the CNN-BiLSTM-AdaBoost algorithm was used to conduct short-term structural displacement prediction and rapid inversion of construction loads, providing early warning for high-formwork support structures. Based on the research results, the main conclusions are as follows:

Based on the research results, the main conclusions are as follows:

(1) Purely data-driven methods show obvious limitations in predicting tower crane displacements under extreme typhoon conditions and tend to underestimate the structural responses of tower cranes.
(2) The hybrid data-model-driven method integrates real-time monitoring data with extreme typhoon simulation results, significantly improving the accuracy and robustness of tower crane displacement prediction under extreme typhoon conditions.

Computer vision-based real-time displacement monitoring of high-formwork support structures:

(1) Illumination intensity: M-DAVIM achieves optimal accuracy within the range of 200–400 lx. Under low illumination or extreme lighting conditions, errors increase significantly but remain within acceptable engineering tolerances. Fog concentration: A GAN-based integrated image restoration module can effectively compensate for image blurring caused by fog and maintains sub-millimeter accuracy even under dense fog conditions. Camera elevation angle: The visual system maintains high accuracy when the elevation angle is within 30°. Target size: When the target size is smaller than 18 pixels, measurement errors increase; therefore, when the M-DAVIM system is used outdoors, the marker size should be greater than 18 pixels. Camera vibration: M-DAVIM can

effectively reduce vibration-induced errors in the frequency range of 5–50 Hz, achieving up to a 92% reduction in error compared with traditional methods.

(2) Taking the monitoring data on May 27, 2023, as an example, the all-day displacement and inclination of the high-formwork support were analyzed. The time-history curves indicate that, due to long-term operational accumulation, different vertical members exhibit different horizontal and vertical displacements; however, all monitored indicators are smaller than the alarm limits specified by the relevant codes, indicating that the structure is safe. The inclination values of the vertical members are far below the inclination alarm thresholds, indicating that vertical member inclination is not the controlling monitoring factor in this project, and greater emphasis should be placed on monitoring horizontal and vertical displacements of the vertical members.

(3) In this chapter, the monitoring and warning thresholds were determined as fixed values according to the Technical Code for Automated Monitoring of Steel Tube Support Frames in Construction (DBJ33/T 1296–2023). However, during actual monitoring, structural conditions continuously change due to factors such as component aging and load variations; therefore, it is necessary to further investigate scientifically reasonable methods for determining and dynamically updating warning thresholds.

Hybrid data-model-driven updating of monitoring and warning thresholds and short-term response prediction for high-formwork support structures:

(1) The proposed framework for determining and updating monitoring and warning thresholds is a dynamic and continuously improving process. Alarm thresholds need to be iteratively updated to accurately match actual structural characteristics. The updated structural buckling failure displacements are all greater than the initial alarm thresholds, indicating that the alarm thresholds are conservatively set and reasonable.

(2) The CNN-BiLSTM-AdaBoost hybrid deep learning model demonstrates strong robustness and high prediction accuracy. This algorithm enables short-term prediction of structural responses and real-time inversion of construction loads, showing its application potential in practical project monitoring.

6.3 Future Research Directions

This study has certain limitations that warrant further improvement. Although semi-rigid connections were adopted in the FEM, the rotational stiffness of joints may vary due to construction quality. Future research should incorporate pre-construction sampling tests to more accurately characterize and model joint rotational stiffness.

References

1. Arcadius Tokognon, C., et al.: Structural health monitoring framework based on internet of things: a survey. IEEE Internet Things J. **4**(3), 619–635 (2017)
2. Hou, K.M., et al.: Trends and Challenges in AIoT/IIoT/IoT Implementation. Sensors **23**(11) (2023)
3. Ali, A.H., Zayed, T., Hussein, M.: Crane safety operations in modular integrated construction. Autom. Constr. **164** (2024)
4. Ali, A.H., et al.: Tower crane safety technologies: a synthesis of academic research and industry insights. Autom. Constr. **163** (2024)
5. Wu, H., et al.: On-site safety inspection of tower cranes: a blockchain-enabled conceptual framework. Saf. Sci. **153** (2022)
6. Zhong, D., et al.: A practical application combining wireless sensor networks and internet of things: safety management system for tower crane groups. Sensors **14**(8), 13794–13814 (2014)
7. Cheng, T., Teizer, J.: Modeling tower crane operator visibility to minimize the risk of limited situational awareness. J. Comput. Civ. Eng. **28**(3) (2014)
8. Luo, X., Leite, F., O'Brien, W.J.: Location-aware sensor data error impact on autonomous crane safety monitoring. J. Comput. Civ. Eng. **29**(4) (2015)
9. Yuan, X., Anumba, C.J., Parfitt, M.K.: Cyber-physical systems for temporary structure monitoring. Autom. Constr. **66**, 1–14 (2016)
10. Chen, S.J., et al.: An intelligent high-formwork support monitoring system. Artif. Intell. Evol. 160–169 (2022)
11. Moon, S., Choi, B., Yang, B.: USN-based data acquisition for increasing safety in the concrete formwork operation. J. Comput. Civ. Eng. **26**(3), 271–281 (2012)
12. Moon, S., Choi, E., Yang, B.: Holistic integration based on USN technology for monitoring safety during concrete placement. Autom. Constr. **57**, 112–119 (2015)
13. Moon, S.: Application of mobile devices in remotely monitoring temporary structures during concrete placement. Procedia Eng. **196**, 128–134 (2017)
14. Feng, D., Feng, M.Q.: Computer vision for SHM of civil infrastructure: from dynamic response measurement to damage detection—a review. Eng. Struct. **156**, 105–117 (2018)
15. Xu, Y., Brownjohn, J.M.W.: Review of machine-vision based methodologies for displacement measurement in civil structures. J. Civ. Struct. Heal. Monit. **8**(1), 91–110 (2017)
16. Xiaowei, Y., Chuanzhi, D.: A review of computer vision-based structural displacement monitoring. China J. Highw. Transp. **32**(11), 21–39 (2019). (In Chinese)
17. Ying, Z., Lixun, Z., Tong, L., et al.: Computer vision-based structural system identification. Chin. Civil Eng. J. **51**(11), 17–23 (2018). (In Chinese)
18. Pan, B., Tian, L., Song, X.: Real-time, non-contact and targetless measurement of vertical deflection of bridges using off-axis digital image correlation. NDT and E Int. **79**, 73–80 (2016)

Q. Li et al., *A Hybrid Data-Model and AI-Driven Approach for Structural Monitoring in Hazardous Construction*, https://doi.org/10.1007/978-981-95-8688-2

19. Feng, D., Feng, M.Q.: Identification of structural stiffness and excitation forces in time domain using noncontact vision-based displacement measurement. J. Sound Vib. **406**, 15–28 (2017)
20. Hsieh, Y.-A., Tsai, Y.J.: Machine learning for crack detection: review and model performance comparison. **34**(5), 04020038 (2020)
21. Cha, Y.J., Choi, W., Büyüköztürk, O.: Deep learning-based crack damage detection using convolutional neural networks. Comput.-Aided Civ. Infrastruct. Eng. **32**(5), 361–378 (2017)
22. Mas, D., et al.: Methods and algorithms for video-based multi-point frequency measuring and mapping. Measurement **85**, 164–174 (2016)
23. Pan, B., et al.: Two-dimensional digital image correlation for in-plane displacement and strain measurement: a review. Meas. Sci. Technol. **20**(6) (2009)
24. Guo, Y.: Risk target identification and tracking methods for traffic construction sites based on computer vision. Harbin Institute of Technology (2022) (In Chinese)
25. Xiong, R., et al.: Onsite video mining for construction hazards identification with visual relationships. Adv. Eng. Inform. **42** (2019)
26. Gao, X., et al.: Structural displacement estimation by a hybrid computer vision approach. Mech. Syst. Signal Process. **204** (2023)
27. Lu, B., Bai, B., Zhao, X.: Vision-based structural displacement measurement under ambient-light changes via deep learning and digital image processing. Measurement **208** (2023)
28. Huang, T., et al.: Spatial displacement tracking of vibrating structure using multiple feature points assisted binocular visual reconstruction. Eng. Appl. Artif. Intell. **126** (2023)
29. Ye, X.W., et al.: Vision-based structural displacement measurement: system performance evaluation and influence factor analysis. Measurement **88**, 372–384 (2016)
30. Zhou, H.F., et al.: Temperature effects on vision measurement system in long-term continuous monitoring of displacement. Renew. Energy **114**, 968–983 (2017)
31. Xu, G., Li, L.: Analysis of safety behavior of prefabricated building workers' hoisting operation based on computer vision. Math. Probl. Eng. **2022**, 1–9 (2022)
32. Sitompul, T.A.: Human-machine interface for remote crane operation: a review. Multimodal Technol. Interact. **6**(6) (2022)
33. Jeong, I., et al.: Vision-based productivity monitoring of tower crane operations during curtain wall installation using a database-free approach **37**(4), 04023015 (2023)
34. Yang, Z., et al.: Safety distance identification for crane drivers based on mask R-CNN. Sensors **19**(12) (2019)
35. Wang, Y., Suzuki, H., Ohtake, Y.: Spatial maps with working area limit line from images of crane's top-view camera. Autom. Constr. **142** (2022)
36. Price, L.C., et al.: Multisensor-driven real-time crane monitoring system for blind lift operations: Lessons learned from a case study. Autom. Constr. **124** (2021)
37. Wang, J., et al.: Vision-based automated recognition and 3D localization framework for tower cranes using far-field cameras. Sensors **23**(10) (2023)
38. Yang, J., et al.: Vision-based tower crane tracking for understanding construction activity **28**(1), 103–112 (2014)
39. Changzheng, C., Bo, P., Jiangtao, R., et al.: Safety monitoring method for high formwork support structures based on visual recognition. Ind. Constr. **54**(02), 65–72 (2024). (In Chinese)
40. Feng, Y., Dai, F., Zhu, H.-H.: Evaluation of feature- and pixel-based methods for deflection measurements in temporary structure monitoring. J. Civ. Struct. Heal. Monit. **5**(5), 615–628 (2015)
41. Jung, Y., Oh, H., Jeong, M.M.: An approach to automated detection of structural failure using chronological image analysis in temporary structures. Int. J. Constr. Manag. **19**(2), 178–185 (2018)
42. Luo, R., et al.: 3D deformation monitoring method for temporary structures based on multi-thread LiDAR point cloud. Measurement **200** (2022)
43. Hangyu, L., Jie, G., Yufei, T., et al.: Displacement monitoring of high formwork support systems based on lightweight multi-object visual perception. J. Build. Struct. **45**(07), 69–79 (2024). (In Chinese)

44. Zhang, Z., Pan, W.: Virtual reality supported interactive tower crane layout planning for high-rise modular integrated construction. Autom. Constr. **130** (2021)
45. He, F., Ong, S.K., Nee, A.Y.C.: An integrated mobile augmented reality digital twin monitoring system. Computers **10**(8) (2021)
46. Shanti, M.Z., et al.: Real-time monitoring of work-at-height safety hazards in construction sites using drones and deep learning. J. Safety Res. **83**, 364–370 (2022)
47. Baduge, S.K., et al.: Artificial intelligence and smart vision for building and construction 4.0: machine and deep learning methods and applications. Autom. Constr. **141** (2022)
48. Mohtasham Moein, M., et al.: Predictive models for concrete properties using machine learning and deep learning approaches: A review. J. Build. Eng. **63** (2023)
49. Zhang, C., Hammad, A.: Improving lifting motion planning and re-planning of cranes with consideration for safety and efficiency. Adv. Eng. Inform. **26**(2), 396–410 (2012)
50. Hyun, H., et al.: Tower crane location optimization for heavy unit lifting in high-rise modular construction. Buildings **11**(3) (2021)
51. Khodabandelu, A., Park, J., Arteaga, C.: Improving multitower crane layout planning by leveraging operational flexibility related to motion paths. J. Manag. Eng. **39**(5) (2023)
52. Lu, Y., et al.: Automated detection of dangerous work zone for crawler crane guided by UAV images via Swin transformer. Autom. Constr. **147** (2023)
53. Wang, Z., et al.: Integrated reinforcement and imitation learning for tower crane lift path planning. Autom. Constr. **165** (2024)
54. Ding, C., et al.: Identifying unsafe behaviors of construction workers through an unsupervised multi-anomaly GAN approach. Autom. Constr. **165** (2024)
55. Sakhakarmi, S., et al.: Automated scaffolding safety analysis: strain feature investigation using support vector machines **47**(8), 921–928 (2020)
56. Jingqiang, S., Sicong, Y., Dongdong, W., et al.: Application of genetic algorithm-optimized BP neural networks in fault diagnosis of tower cranes. Hoisting Conveying Mach. **04**, 61–64 (2012). (In Chinese)
57. Jiang, W., Ding, L.: Unsafe hoisting behavior recognition for tower crane based on transfer learning. Autom. Constr. **160** (2024)
58. Yin, J., et al.: Optimization of service scheduling problem for overlapping tower cranes with cooperative coevolutionary genetic algorithm. Eng. Constr. Archit. Manag. **31**(3), 1348–1369 (2022)
59. Kim, J., et al.: Point cloud registration considering safety nets during scaffold installation using sensor fusion and deep learning. Autom. Constr. **159** (2024)
60. Shen, T., Nagai, Y., Gao, C.: Design of building construction safety prediction model based on optimized BP neural network algorithm. Soft. Comput. **24**(11), 7839–7850 (2019)
61. Smoczek, J., Szpytko, J.: Evolutionary algorithm-based design of a fuzzy TBF predictive model and TSK fuzzy anti-sway crane control system. Eng. Appl. Artif. Intell. **28**, 190–200 (2014)
62. Jianrong, Z., Wei, Z., Nannan, X., et al.: Safety accident prediction and cause analysis of tower cranes based on random forest algorithm. Saf. Environ. Eng. **28**(05), 36–42 (2021). (In Chinese)
63. Yanbin, X., et al.: Research on optimization of crane fault predictive control system based on data mining. Nonlinear Eng. **12**(1) (2023)
64. Li, Q., et al.: Machine learning-based prediction of dynamic responses of a tower crane under strong coastal winds. J. Mar. Sci. Eng. **11**(4) (2023)
65. Zhang, M., Ge, S.: Vision and trajectory-based dynamic collision prewarning mechanism for tower cranes **148**(7), 04022057 (2022)
66. Wei, C.-C.: Collapse warning system using LSTM neural networks for construction disaster prevention in extreme wind weather. J. Civ. Eng. Manag. **27**(4), 230–245 (2021)
67. Han, J., Bixiong, Z., Yicheng, L., et al.: Safety analysis of tower cranes based on BP neural networks optimized by the cuckoo search algorithm. Ind. Constr. **53**(S2), 790–793 (2023). (In Chinese)

68. Tamascelli, N., et al.: Learning from major accidents: a machine learning approach. Comput. Chem. Eng. **162** (2022)
69. Tang, S., et al.: A review of building information modeling (BIM) and the internet of things (IoT) devices integration: present status and future trends. Autom. Constr. **101**, 127–139 (2019)
70. Hu, X., Assaad, R.H.: A BIM-enabled digital twin framework for real-time indoor environment monitoring and visualization by integrating autonomous robotics, LiDAR-based 3D mobile mapping, IoT sensing, and indoor positioning technologies. J. Build. Eng. **86** (2024)
71. Tak, A.N., et al.: BIM-based 4D mobile crane simulation and onsite operation management. Autom. Constr. **128** (2021)
72. Cheng, J.C.P., et al.: Data-driven predictive maintenance planning framework for MEP components based on BIM and IoT using machine learning algorithms. Autom. Constr. **112** (2020)
73. Pan, Y., Zhang, L.: A BIM-data mining integrated digital twin framework for advanced project management. Autom. Constr. **124** (2021)
74. Zhang, J., et al.: Digital twins for construction sites: concepts, LoD definition, and applications. J. Manag. Eng. **38**(2) (2022)
75. Gao, K., Xu, X., Jiao, S.: Prediction and visualization analysis of drilling energy consumption based on mechanism and data hybrid drive. Energy **261** (2022)
76. Zhenyu, H., Yiming, W., Dazhen, Z., et al.: Hybrid mechanism-data-driven modeling method for urban drainage pipe networks. Water Wastewater Eng. **60**(11), 149–159 (2024). (In Chinese)
77. Changxin, C., Kewei, W., Liqian, W., et al.: Dynamic modeling of hunting dampers based on a mechanism-data hybrid-driven approach. Modern Urban Transit **01**, 72–80 (2025). (In Chinese)
78. Yi, Z., Shouquan, S., Siyang, L., et al.: Mechanism-data hybrid-driven short-process power model for steel production enterprises. Proc. Chin. Soc. Electr. Eng. **45**(18), 7098–7110 (2025). (In Chinese)
79. Johns, B., Abdi, E., Arashpour, M.: Dynamical modelling of boom tower crane rigging systems: model selection for construction. Arch. Civ. Mech. Eng. **23**(3) (2023)
80. Ebrahimian, H., et al.: Estimation of soil-structure model parameters for the millikan library building using a sequential Bayesian finite element model updating technique. Buildings **13**(1) (2022)
81. Roohi, M., Hernandez, E.M., Rosowsky, D.: Nonlinear seismic response reconstruction and performance assessment of instrumented wood-frame buildings-Validation using NEESWood Capstone full-scale tests. Struct. Control. Health Monit. **26**(9) (2019)
82. Chen, W., et al.: Wind-induced tower crane vibration and safety evaluation. J. Low Freq. Noise, Vib. Act. Control. **39**(2), 297–312 (2019)
83. Ereiz, S., Duvnjak, I., Fernando Jiménez-Alonso, J.: Review of finite element model updating methods for structural applications. Structures **41**, 684–723 (2022)
84. Ghahari, F., et al.: A hybrid model-data method for seismic response reconstruction of instrumented buildings. Earthq. Spectra **40**(2), 1235–1268 (2024)
85. Wang, Y.: Simulation study on ball-end milling tool wear during TC4 milling processes. Harbin University of Science and Technology (2020) (In Chinese)
86. Xiaogang, Q., Xiaokang, Z., Songyan, J., et al.: *Numerical simulation of remaining service life of bridge cranes based on fracture mechanics*. J. Saf. Environ. **22**(03), 1284–1290 (2022). (In Chinese)
87. Wang, C., et al.: System reliability-based limit state design of support scaffolding systems. Eng. Struct. **216** (2020)
88. Cho, C., et al.: Data-driven monitoring system for preventing the collapse of scaffolding structures. J. Constr. Eng. Manag. **144**(8) (2018)
89. Sun, L., et al.: Hybrid monitoring methodology: A model-data integrated digital twin framework for structural health monitoring and full-field virtual sensing. Adv. Eng. Inform. **60** (2024)

90. Jiang, W., Ding, L., Zhou, C.: Digital twin: Stability analysis for tower crane hoisting safety with a scale model. Autom. Constr. **138** (2022)
91. Liu, Z., et al.: Digital twin-based safety evaluation of prestressed steel structure. Adv. Civ. Eng. **2020**(1) (2020)
92. Liu, Z., et al.: Intelligent prediction method for operation and maintenance safety of prestressed steel structure based on digital twin technology. Adv. Civ. Eng. **2021**(1) (2021)
93. Shan, L., et al.: CNN-BiLSTM hybrid neural networks with attention mechanism for well log prediction. J. Pet. Sci. Eng. **205** (2021)
94. Ma, T., et al.: Horizontal in situ stresses prediction using a CNN-BiLSTM-attention hybrid neural network. Geomech. Geophys. Geo-Energy Geo-Resour. **8**(5) (2022)
95. Wei, Z., et al.: A method for sound speed profile prediction based on CNN-BiLSTM-attention network. J. Mar. Sci. Eng. **12**(3) (2024)
96. Xu, J., et al.: A novel and robust data anomaly detection framework using LAL-AdaBoost for structural health monitoring. J. Civ. Struct. Heal. Monit. **12**(2), 305–321 (2022)
97. Huang, M., et al.: Typhoon wind hazard estimation by full-track simulation with various wind intensity models. J. Wind. Eng. Ind. Aerodyn. **218** (2021)
98. Vickery, P.J., Wadhera, D.: Statistical models of Holland pressure profile parameter and radius to maximum winds of hurricanes from flight-level pressure and h*wind data. J. Appl. Meteorol. Climatol. **47**(10), 2497–2517 (2008)
99. Li, Q., et al.: Typhoon-induced fragility analysis of transmission tower in ningbo area considering the effect of long-term corrosion. Appl. Sci. **12**(9) (2022)
100. Li, Q., et al.: Typhoon loss assessment in rural housing in Ningbo based on township-level resolution. Appl. Sci. **12**(7) (2022)
101. Busca, G., et al.: Vibration monitoring of multiple bridge points by means of a unique vision-based measuring system. Exp. Mech. **54**(2), 255–271 (2013)
102. Liu, X., et al.: Videogrammetric technique for three-dimensional structural progressive collapse measurement. Measurement **63**, 87–99 (2015)
103. Huang, M., et al.: A deep learning augmented vision-based method for measuring dynamic displacements of structures in harsh environments. J. Wind. Eng. Ind. Aerodyn. **217** (2021)
104. Redmon, J., et al.: You only look once: unified, real-time object detection. In: 2016 IEEE Conference on Computer Vision and Pattern Recognition (CVPR) (2016)
105. Huang, M., Zhang, B., Lou, W.: A computer vision-based vibration measurement method for wind tunnel tests of high-rise buildings. J. Wind Eng. Ind. Aerodyn. **182**, 222–234 (2018)
106. Li, Q., et al.: M-DAVIM-based vibration monitoring and factor analysis for tower crane structures under complex construction environments. Structures **82** (2025)

MIX
Papier aus verantwortungsvollen Quellen
Paper from responsible sources
FSC® C105338

If you have any concerns about our products,
you can contact us on
ProductSafety@springernature.com

In case Publisher is established outside the EU,
the EU authorized representative is:
Springer Nature Customer Service Center GmbH
Europaplatz 3, 69115 Heidelberg, Germany

Printed by Libri Plureos GmbH
in Hamburg, Germany